Writing Reports
to Get Results

Books of Related Interest from the IEEE Press

The Essence of Technical Communication for Engineers: Writing, Presentation, and Meeting Skills
Herbert L. Hirsch
2000   Softcover   89 pp.   IEEE Order No. PP5854   ISBN 0-7803-4738-2

The Woman's Guide to Navigating the Ph.D. in Engineering and Science
Barbara B. Lazarus, Lisa M. Ritter, and Susan A. Ambrose
2001   Softcover   105 pp.   IEEE Order No. PP5883   ISBN 0-7803-6037-0

Hargrave's Communications Dictionary
Frank Hargrave
2001   Hardcover   917 pp.   IEEE Order No. PC5869   ISBN 0-7803-6020-6

Writing Reports to Get Results

*Quick, Effective Results
Using the Pyramid Method*

Third Edition

Ron S. Blicq
Lisa A. Moretto
RGI International

IEEE Professional Communication Society, Sponsor

IEEE Press

WILEY-INTERSCIENCE

A JOHN WILEY & SONS, INC., PUBLICATION
New York • Chichester • Weinheim • Brisbane • Singapore • Toronto

This text is printed on acid-free paper. ∞

Copyright © 2001 by the Institute of Electrical and Electronics Engineers, Inc. All rights reserved.

For ordering and customer service, call 1-800-CALL-WILEY.

Library of Congress Cataloging-in-Publication Data is available.

ISBN 0-471-14342-1

10 9 8 7 6 5 4 3

6511213

Contents

Preface

We have prepared these guidelines as an easy-to-consult reference handbook, designed especially for people who work in a business or technical environment and have to write reports. Its tailor-made writing plans can help you, as a manager, business administrator, researcher, supervisor, engineer, scientist, technician, computer specialist, or student, start writing more readily and continue writing more easily.

The writing plans cover the three general categories of reports written in business, government, and industry. Short reports include informal incident, field trip, job progress, project completion, and inspection reports; semiformal reports comprise laboratory reports and medium-length investigation and evaluation reports; and formal reports cover analytical and feasibility studies, as well as major investigations. There are also writing plans for three types of proposals, from single-page suggestions to full-length formal presentations.

All of the writing plans are based on a unique modular method of report organization called the pyramid method, which is described in Chapter 2. This chapter will help you identify the most important information you have to convey and focus your readers' attention on it. The pyramid method then groups the remaining information into compartments that develop your case logically and coherently.

For each type of report, the guidelines provide

- an individual writing plan,

- detailed instructions for using the writing plan,

- a model report (in some cases there are two examples), and

- comments on how the writer has used the suggested writing plan to shape his or her report.

A writing techniques section at the end of the handbook provides useful suggestions for "sprucing up" the appearance of your reports and getting better mileage from your words. It also describes how to construct a list of references or a bibliography; how to present numbers, abbreviations, and metric (SI) symbols; how to prepare illustrations for insertion within a report's narrative; and how to work collaboratively as one of several members engaged in writing a comprehensive report or proposal.

RB and LM

PART 1

A Practical Approach to Report Writing

CHAPTER 1

How to Use These Guidelines

There are two ways you can use these guidelines: you can read them right through from start to finish, or you can read only the parts that apply to the kind of report writing you do. If you are a busy person, you are more likely to read selectively.

If you choose to dip into sections of the book, we recommend you follow this reading plan:

1. *Be sure to read Chapter 2 first.* This is a particularly important chapter because it describes the basic structure on which all the reports in Chapters 3 through 8 are modelled.

2. From the Table of Contents identify which report types listed in Parts 2, 3, and 4 (Chapters 3 through 8) you write now. Also identify any report types you think you might have to write over the next 12 months.

3. Turn to each of the reports you have identified and then:

   • Read the introductory remarks and recommended writing plan.

   • Read the model report. You will find most model reports are printed on right-hand pages, and most comments about the reports are printed on the facing left-hand pages. We recommend you first read the model report right through once and resist the temptation to glance across to the cross-referenced comments on the facing page(s). This will give you a better "feel" for the report.

   • Read the comments on the facing page(s) and cross-reference them to the report.

   Note: For some reports you write, you will find an exact writing plan to use and a comparable model to follow in the guidelines. For others, you may have to search for a writing plan that approximates your needs, and then adapt it to fit your particular situation.

4. Read Part 5 (Chapters 9 through 15). These chapters contain "how to" suggestions on report shape, appearance, language, and writing style, and so act as a reference section which you can consult at any time.

5. When you have the time or the need, turn to the other reports and review the pyramids and the examples.

The individual writing plans illustrated here have been tested and used thousands of times and are known to work well. But bear in mind that they are only *suggestions* for organizing each report. They are not hard-and-fast rules, and you can alter them to suit both your needs and those of your audience—the person or people for whom you write each report.

CHAPTER 2

The Report Writer's Pyramid

If we were to ask you what you find most difficult about report writing, would one of these be your answer?

- *Getting started.*
- *Organizing the information: arranging it in the proper order.*
- *The writing: getting the right words down on paper the first time.*

We can ask the same question of any group of business or technical report writers and always hear the same answers. And often those who say getting started also mention one of the other answers.

The ideas presented in this book will help remove some of the drudgery from report writing. They will show you how to get started, organize your thoughts, and write simply and easily. This chapter provides you with basic guidelines. Subsequent chapters demonstrate how you can apply the guidelines to various situations.

Getting Started

Dave Kowalchuk has spent two weeks examining his company's methods for ordering, receiving, storing, and issuing parts for the electronic equipment the company services. He has discovered that the inventory control system is inefficient, and has investigated alternative methods and devised a better system. Now he is ready to write a report describing his findings and suggestions.

But Dave is having trouble getting started. When he sits down to write, he just can't seem to find the right words. He writes a few sentences, and sometimes several paragraphs, yet each time he discards them. He is frustrated because he is unable to bring his message into focus.

Dave's problem is not unusual. It stems from a simple omission: he has neglected to give sufficient thought either to his reader or to the message he has to convey. He needs to make three critical decisions *before* he picks up his pen or places his fingers on the computer keyboard. He should ask himself:

1. *Who is my reader?*
2. *What do I most want to tell that reader?*
3. *What will the reader do with this information?*

Identifying the Reader

If you are writing a memo report to your manager, you will know immediately to whom you are writing (although you may have to give some thought to other possible readers, if your manager is likely to circulate your memo). But if your report will have a wider readership—as Dave's may well have—then you must decide who is to be your primary reader and write for that particular person. Trying to write for a broad range of readers can be as difficult as trying to write with no particular reader in mind. In both cases you will have no focal point for your message. And without a properly defined focal point your message may be fuzzy.

How can you identify the primary reader? It is the person (or people) who will probably use or act upon the information you provide. You need not know the person by name, although it is useful if you do because then you will have a precise focal point. But you should at least know the type of person who will use your information and be able to identify the position he or she holds.

Yet simply knowing your reader is not enough. You need to carry the identification process one step further by answering four more questions:

QUESTION 1: *What does the reader want, expect, or need to hear from me?*
You have to decide whether your reader will want a simple statement of facts or a detailed explanation of circumstances and events. You also have to consider whether the reader needs to know how certain facts were derived.

QUESTION 2: *How much does the reader know already?*
The answer to this question will provide you with a starting point for your report, since there is no need to repeat information the reader already knows. (But note that your answer may also be influenced by the answer to question 4.)

QUESTION 3: *What effect do I want my report to have on the reader?*
You have to decide whether the purpose of your report is to inform or to persuade. In an informative report you simply relate the necessary facts, and then you stop. In a persuasive report you have to convince the reader to act or react, which can mean simply approving a plan you propose, or ordering materials or equipment on your behalf, or authorizing a change in policies and procedures.

QUESTION 4: *Are other people likely to read my report?*
You have to consider the route your report takes before and after it reaches your reader, and to whom you may send copies. If the report will pass through other people's hands, then you must consider how much additional information you will have to insert to satisfy their curiosity. (At the same time you must not let your desire to satisfy additional readers deflect you from focusing on the primary reader's needs and expectations.)

In the situation described earlier, Dave Kowalchuk decides his primary reader is Maria Pavanno, who is Manager of Purchasing and Supply. He also recognizes that Maria may circulate his report to other managers and particularly to the Vice President of the division. He also needs to consider these secondary readers.

Identifying the Message

Now that Dave has his primary and secondary readers clearly in mind, he has to make a second decision. This time he has to answer a single question:

What do I most want to tell my primary reader?

Dave must examine the results of his investigation and decide which results will be most useful to Maria Pavanno. His aim should be to find key information that will spark Maria's interest so she will want to know more. For example, would she *most* want to know that:

1. The company's supply system is out of date and inefficient?
2. Other businesses Dave has investigated have better supply systems?
3. There are several ways the company's supply system can be improved?
4. Improvements to the company's supply system will increase efficiency?
5. Changes to the supply system will save time and money?

Although all these points are valid, Dave reasons that Maria will be most interested in knowing how to save the department time and money. As increased efficiency is the key to these savings, he decides to combine points 4 and 5 into a single message. So he writes

Improvements to our inventory control system will increase efficiency and save time and money.

This becomes his *Main Message*—the information he most wants to convey to his reader, Maria Pavanno.

When you have identified both your primary reader and your Main Message, write them in bold letters on a separate sheet of paper and keep the sheet in front of you as you write. This will give you a constant reminder that you are writing for a particular person and have a specific purpose in mind.

Using the Pyramid Method

If you were to ask any group of managers what single piece of advice they would give new report writers, the two replies you would hear more than any others are:

- Tell me right away what I most need to know.

- Draw my attention to the results. Don't bury them so I have to hunt for them.

You can meet both of these requirements if you use the pyramid method to organize your reports. The pyramid method emphasizes the most important information by bringing it right up front, *where it will be seen.*

Figure 2–1   The report writer's pyramid.

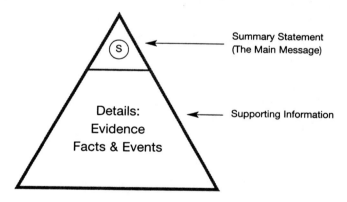

As its name implies, the pyramid method suggests that you organize your reports in pyramid form, as shown in Figure 2–1. The essential information (what the reader most needs to know) sits at the top of the pyramid, where it is supported by a strong base of facts and details.

This concept is not new. Journalists (whom we often call "reporters") have used it for decades, and in recent years experienced business and technical report writers have adopted it, because it offers them the most efficient way to communicate information.

The pyramid is used as the basic framework for organizing every type of report, although the compartments may be relabelled and expanded to suit varying situations. When you have become accustomed to using it, you will find that you automatically think "pyramid-style" every time you write.

Focusing the Message

Many beginning report writers find the pyramid method difficult to accept, because it seems to contradict what they have previously been taught. Throughout high school and into college or university they have probably been told to write using the "climactic" method. Writing climactically means developing a topic carefully, logically, and sequentially, so that the narrative leads systematically up to the main point. It is the ideal way to write an essay, short story, or mystery novel, in which the main point needs to be at the end of the piece of writing. But it does not meet the needs of business and industry, where readers want to find the Main Message at the beginning.

The pyramid method suggests that you identify the most important information in the Details section and then summarize it into a short Summary Statement, which you place at the front or top of your report. In this way you focus your readers' attention onto your Main Message.

If you have been accustomed to writing climactically, you may feel uncomfortable jumping straight into the Main Message without first gently leading up to it. To start, you can borrow a technique used by newspaper reporters.

Turn to the front page of your daily newspaper and read the first few paragraphs of each article. You will find that every article is structured the same way:

1. It has a headline, which is not really part of the article. (Normally headlines are written by editors, not the newspaper reporters who write the articles.)

2. Its opening paragraph very briefly gives you the main information—usually what has happened and, sometimes, the outcome. For example:

> Maps distributed in Washington yesterday show that the 77.5-ton Monitor unmanned space station will pass directly over Buffalo, New York on three occasions during its final day in orbit. It is expected to re-enter the earth's atmosphere on Wednesday July 11, plus or minus one day.

   This opening paragraph is the article's Main Message and is equivalent to the Summary Statement at the front of a short report. It is intended to grab the readers' attention and create enough interest to keep them reading.

3. Its remaining paragraphs expand the Main Message by providing details such as facts, events, names of places and people, dates and times, and statements by persons the reporter has interviewed. It is equivalent to the Details section of a report.

What you cannot see at the front of each article are six "hidden words," which newspaper reporters were taught to use every time they start writing. First they wrote

I want to tell you that ...

Then they finished the sentence with their Main Message (what they most wanted to tell their readers). For example:

> I want to tell you that ... there are larvae in the city's water supply, but local authorities say they don't pose a threat to public health.

Finally, they removed the six words "I want to tell you that ..." (which is why they are known as hidden words), so that the remaining words became the article's opening sentence.

You can see how this is done if we restore the six hidden words to the front of the opening paragraph describing the orbiting space station:

> I want to tell you that ... maps distributed in Washington yesterday show that the 77.5-ton Monitor unmanned space station will pass directly over Buffalo, New York on three occasions during its final day in orbit.

Similarly, you can insert the six hidden words before the opening sentence of every article on the front page of your daily newspaper. The six hidden words force you to identify who you are writing to, what you want to tell that person (or these people), and to make the message come directly from you.

You can use the hidden words technique to help you start every report you write. Just follow these steps:

1. Identify your reader.

2. Decide what you *most* want to tell your reader.

3. Write down the six words *I want to tell you that ...*

4. Complete the sentence by writing what you decided to tell your reader in step 2. This is your Main Message.

5. Delete the six hidden words of step 3.

Here is how Dave Kowalchuk used these five steps to start his report on the company's supply system:

1. He identified his primary reader as Maria Pavanno, Manager of Purchasing and Supply.

2. He decided he most wanted to tell Maria that the department can save time and money by improving its inventory control system.

3. He wrote: "I want to tell you that ..."

4. He finished the sentence by writing: "... improvements to our inventory control system will increase efficiency and save time and money."

5. He deleted the six words he had written in step 3.

Step 4 became the opening sentence in Dave's report; that is, he used it as his Summary Statement (or Main Message). But when Dave examined the words more closely he realized that, although what he had written was accurate, as an opening statement it was too abrupt. He remembered that a Summary Statement must not only inform but also create interest and encourage the reader to continue reading. So he rearranged his information and inserted additional words to soften the abruptness. At the same time he took great care not to lose sight of his original message. After several attempts he wrote

> (*I want to tell you that ...*) My examination of our inventory control system shows we can increase departmental efficiency, save time, and reduce costs by improving our methods for ordering, storing, and issuing stock.

We suggest that you, like Dave Kowalchuk, use the "hidden words" method every time you have to write a report. It will help you start more easily and ensure that you focus your readers' attention immediately onto the most important information.

Developing the Details

Because the Summary Statement of a report brings readers face to face with important, sometimes critical, and occasionally controversial information, it immediately triggers questions in their minds. Your responsibility is to anticipate these questions and answer them as quickly and efficiently as you can. You do this in the Details section, which amplifies and provides supportive evidence for the main message in your Summary Statement.

There are six questions a reader may ask: *Who?*, *Why?*, *Where?*, *When?*, *What?*, and *How?* (See Figure 2–2.) But first you have to identify which of these questions your readers would most likely ask. Say to yourself:

> *If I were the intended reader, which questions would I ask after I had read **only** the Summary Statement?*

(When answering this question, remember that your readers will not know the subject nearly as well as you do.)

Dave Kowalchuk, for example, might say to himself: "What questions will Maria Pavanno be likely to ask immediately after she has read my Summary Statement?" (The Summary Statement is in bold type above) Dave would probably come up with the following questions:

- **Why** *(is it necessary to increase efficiency)*?
- **How** *(can we improve efficiency)*?
- **What** *(will be the effect or result of improved efficiency)*?

If he thinks that Maria wants a very detailed report, Dave might also ask these questions:

Figure 2–2   The base of the report writer's pyramid answers readers' questions.

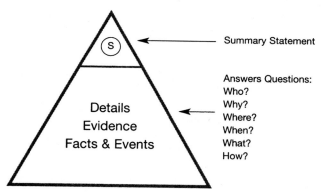

- **When** *(should the improvements be implemented)?*
- **Who** *(will be affected by them)?*
- **How** *(much will it cost)?*

He would omit the question *where?* because for this report it does not need to be answered.

Because Dave recognizes that Maria may circulate his report to other managers or to the vice president of the division, he uses a language that is appropriate to these secondary audiences. For example, he includes information to help managers understand the situation and make decisions but leaves out many of the technical details and findings. They would be used to justify or prove his recommendation and could be presented as attachments.

Now let's examine how another writer—Bev Hubka—used these questions to develop the Details section of a short inspection report. Recently, Bev drove to a warehouse to determine the condition of some new equipment damaged in a traffic accident and found that most of it was beyond repair. In the Summary Statement of the report Bev told her readers what they most needed to know:

Main Message: *(1 want to tell you that ...)* Our inspection shows that only three of the 16 computers in Calvin Computer Systems' shipment No. 367 can be repaired. The remainder will have to be scrapped.

To assemble facts for the Details section, Bev jotted down notes in answer to the six questions her readers might ask after they had read her Summary Statement:

Who *(was involved)?*
Fran Derwood and Bev Hubka
Why *(were you involved)?*
We had to inspect damaged computers.
(Authority: Arlington Insurance Corporation)

Where *(did you go)?*
> To Hillsborough Storage warehouse

When *(did this happen)?*
> On June 18

What *(did you find out)?*
> Three repairable computers, 13 damaged beyond repair

How *(were they damaged)?*
> In a semitrailer involved in a highway accident

Finally, Bev took these bare facts and shaped and expanded them into two Detail paragraphs.

Details:

Why?	We were requested by Arlington Insurance Corporation to examine the condition of 16 CANFRED computers manufactured by Calvin Computer Systems of Austin, Texas. They were damaged when the semitrailer in which they were shipped overturned and burned on a curve near Jackson, Mississippi, on June 11. Fran Derwood and I drove to Jackson on June 18, where we were met by Arlington Insurance Corporation representative Kevin Cairns. He escorted us to the Hillsborough Storage warehouse.
How?	
Who?	
Where?	
When?	
What?	We found that the fire that resulted from the accident has irreparably damaged 13 computers. Three others suffered smoke damage but seem to be electronically sound. They carry serial numbers 106287, 106291, and 106294. We estimate that these computers will cost an average of $350.00 each to repair, for a total repair cost of $1050.00.

The pyramid method can help you organize random bits of information, just as it has helped Bev Hubka. And, because it helps you eliminate unessential information, it will also shorten the reports you write. But it is not meant to be a rigid method for organizing details. The six basic questions are intended solely as a guide, and should be used flexibly. For example:

- The questions do not have to be answered in any particular order. You can arrange the answers in any sequence you like, balancing your personal preference against the reader's needs and the most suitable way to present your information.

- Only the appropriate questions need to be answered (i.e. the questions that are pertinent to each particular reporting situation).

- The first four questions in the list (*who?, why?, where?,* and *when?*) require fairly straightforward answers. The last two questions (*what?* and *how?*) can have widely varying answers, depending on the event or situation you are reporting. Here you explain what has happened, how it happened, what needs to be done, and possibly how best to resolve the problem or improve the condition. Consequently there is ample scope for originality and ingenuity on your part.

Expanding the Details Section

The pyramid method provides the basic structure for all reports, regardless of their length. It can be used for one-paragraph email messages, one-page memos, and 100-page reports. In every case the report opens with a Summary (or Summary Statement), which presents a synopsis of the main information to be conveyed (that is, the Main Message). The Summary is followed by a longer section containing factual details, which support and amplify the initial statement.

Bev Hubka's report describing the inspection of damaged computers is a typical short report structured using the simple two-part pyramid. For reports of greater length or complexity, however, the Details section at the base of the pyramid needs to be developed further. This development is obtained by expanding the Details section into three basic compartments of information:

- A **Background** compartment, which describes the circumstances leading up to the situation or event. (It answers the questions: *who?, why?, where?,* and *when?*)

- A **Facts & Events** compartment, which describes in detail what happened, or what you found out during your project. (It answers the last two questions: *what?* and *how?*)

- An **Outcome** compartment, which describes the results of the event or project, and sometimes suggests what action needs to be taken. (It also can answer the questions *what?* and *how?*)

The compartments are shown within the pyramid in Figure 2–3 and are identified beside the appropriate parts of the short report in Figure 2–4.

Figure 2–3   The main compartments of a report.

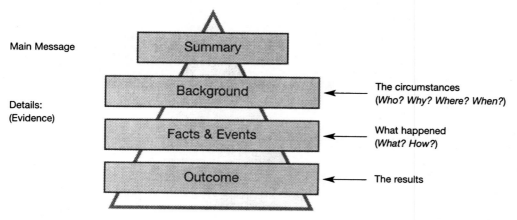

Figure 2–4   A short report structured pyramid style. The main report writing compartments are identified on the left side of the report.

To: Dawn Schwenck, Marketing Manager
         Head Office
From: Karen Peebles, Manager, Store No. 6

Summary

(I want to tell you that ...) The installation instructions included with version 4.2 of the Vancourt 4Tell software program contain errors that must be corrected before we sell any further copies.

Background
Who?
When?
Why?

Version 4.2 is the newly released Windows application of this popular forecasting program and is an upgrade of the earlier version 3.5 that sold so well. We received our first shipment of version 4.2 on October 12 and sold nine copies in the first five days. However, today we received four complaints from customers that the program was impossible to install.

Facts &
Events
What?
How?

I tried installing the software on one of our PCs and discovered that the software instructions have two major and several minor errors that inhibit installation. For example:

- Step 18 tells the installer to press the ALT and F11 keys *simultaneously*, when they should be pressed *consecutively*, which causes the system to shut down completely. The computer then has to be rebooted and the installation restarted.

- Step 23 tells the installer to delete all version 3.5 files, yet three of the previous files must be retained because they carry instructions for converting files from the earlier version.

Outcome
What?
How?

I have attached an addendum to each set of instructions in the 15 software packages we have in stock and have faxed a copy to the manufacturer (Vancourt Software Incorporated of Boston, MA). I will also forewarn our other stores of the error and fax a copy of the addendum to them.

The four compartments

Summary
Background
Facts & Events
Outcome

provide the basic framework for every report you are likely to write. You will be able to identify them in every report described in Chapters 3 through 8, although often you will find the compartments are relabeled to suit particular situations. In longer reports the compartments are also subdivided to accommodate more information and to improve internal organization. These subdivisions occur mostly in the Facts & Events compartment.

In a long report, the different sections within the report can also be structured as individual pyramids, so that smaller pyramids are nested within the overall pyramid.

PART 2

Informal Reports

CHAPTER 3

Incident, Field Trip, and Inspection Reports

Before reading this chapter you should be familiar with the concepts discussed in Chapter 2. The reports in this chapter are short, each containing one to three pages of narrative and occasionally attachments such as drawings, photographs, and calculations.

For every report described here—and in subsequent chapters—we provide

- a rationale and instruction for writing the report,
- an illustration of its pyramid structure, with a definition for each writing compartment,
- a model report, printed on a right-hand page, and
- comments on how the report has been written, printed on the left-hand page facing the model report.

Short reports such as these are often written as memorandums, sometimes as letters, and occasionally as semiformal reports with a title at the top of the page. They may be transmitted using electronic mail. For examples of typical formats, see Chapter 9.

Incident Reports

An incident report (sometimes called an occurrence or accident report) describes an event that has happened, explains how and why it occurred, and indicates what effect the event had and what has been done about it. It may also suggest that corrective action be taken, or what should be done to prevent the event from recurring.

The writing compartments are similar to those of the basic report writer's pyramid and are shown in Figure 3–1.

- The **Summary** compartment identifies the main message for the reader. Chapter 2 describes how to write an effective and attention-getting summary.

- The **Background** compartment answers the questions *Who?*, *Why?*, *Where?*, and *When?*

- The **Facts & Events** and **Outcome** compartments answer the questions *What?* and *How?*

Figure 3–1   The base of the report writer's pyramid answers readers' questions.

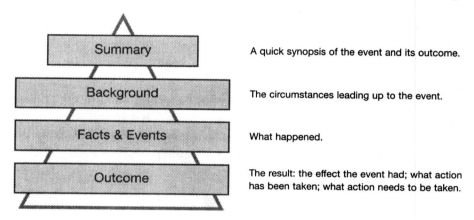

Summary	A quick synopsis of the event and its outcome.
Background	The circumstances leading up to the event.
Facts & Events	What happened.
Outcome	The result: the effect the event had; what action has been taken; what action needs to be taken.

The depth of detail provided in each compartment depends on the importance of the event and how much the reader wants or needs to know. For example, if you were informing your company's accountant of an unexpected expense that caused you to overrun your project budget by $220, you would write just a brief report. But if you were describing an accident that hospitalized two employees and cost $30,000 in repair work, you would be expected to write a detailed report that describes the circumstances and the corrective action that was taken.

The comments on page 20 identify the four writing compartments in the short incident report on page 21.

Incident Report
Reporting a Project Delay
Comments

In the report on the opposite page, project engineer Mark Kagle is describing a technical problem that affected completion of a project.

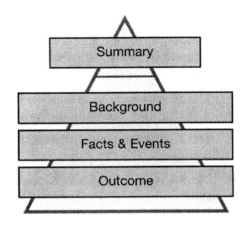

(1) In his **Summary Statement** Mark explains briefly what happened and the effect the event had. The Summary Statement should always have two parts: briefly what happened and what was the result.
(Notice that the hidden words *I want to tell you that* ... can be placed in front of this paragraph *and* all the other paragraphs. This is not unusual for a short report.)

(2) In his **Background** compartment Mark answers the two "background" questions his reader is likely to ask:

> *What* tests? (Data transmission over fiberoptic cable)
> *Who* for? (Waverly Power Commission)

(3) Mark's **Event** compartment describes exactly what happened and how the problem occurred (it answers the questions *Why?*, *What?*, *When?*, and *How?*).

(4) In his **Outcome** compartment Mark answers the question *What?* (*What* was the result of the incident, and *what* have you done about it?)

INTEROFFICE MEMORANDUM

TO: Dwight Murray   DATE: October 13, 2002

FROM: Mark Kagle   SUBJECT: Effect of Power Outage on Project Taurus

(1) An electrical blackout last night interrupted the Project Taurus data transmission tests. This will increase project costs by $2380 and delay the project completion date a further 48 hours.

(2) The tests require continuous transmission of data at 56,000 baud for 24-hour periods over fiberoptic transmission lines ranging in length from 5.6 to 224.3 miles. The tests are being performed for the Waverly Power Commission under contract WTS1771, and were to be completed by October 15, 2002.

(3) The power outage was caused by lightning that disabled three transformers at Wickens Peak power station at 21:17 on October 12. The outage lasted for 3 hours 18 minutes. The lightning strike also created a momentary power surge which damaged two Vancourt 1800 computers we were using for the tests.

(4) I have arranged for two replacement computers to be delivered tomorrow morning, October 14, and will be bringing in additional shift technicians—both here and at the eight receiving points—to work the nights of October 15 and 16. I have also informed Waverly Power Commission that Project Taurus will not be complete until October 17.

Field Trip Reports

Trip reports are written whenever people leave their usual place of work to do something elsewhere. Their reports can cover many kinds of events, such as

- installation or modification of equipment,

- assistance on a field project,

- attendance at a conference, seminar, or workshop,

- repairs to a client's equipment or field instruments, or

- evaluation of another firm's buildings, facilities, or methods.

Whatever the circumstances, the writing compartments for a field trip report are essentially the same as those for the basic report writing pyramid described in Chapter 2. The Event compartment, however, is relabeled Trip Activities, as shown in Figure 3–2.

These compartments generally contain the following information:

- The **Background** compartment describes the purpose of the trip, mentions on whose authority it was taken, and lists circumstantial details such as the names of people involved, dates, and locations.

- The **Trip Activities** compartment describes what was done. Often, it can be broken down into four subcompartments:

  1. What the report writer set out to do.
  2. What was actually done.
  3. What could not be done, and why.
  4. What else was done.

Figure 3–2   Writing compartments for a field trip report.

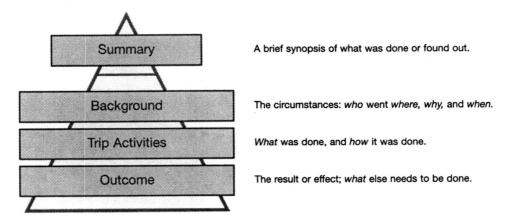

Summary	A brief synopsis of what was done or found out.
Background	The circumstances: *who* went *where, why,* and *when.*
Trip Activities	*What* was done, and *how* it was done.
Outcome	The result or effect; *what* else needs to be done.

The fourth subcompartment is necessary because people on field trips often find themselves doing things beyond the purpose of their assignment. For example, a technician sent to repair a defective diesel power unit at a remote radar site may be asked by the on-site staff to look at a second unit that is "running rough," and spend six hours adjusting its timing cycle. The time spent on this additional work must be accounted for and the work must be described in the trip report.

- The **Outcome** compartment sums up the results of the trip and, if further work still needs to be done or follow-up action should be taken, suggests what is necessary and even how and by whom it should be accomplished.

Many trip reports are short and simply follow the compartment arrangement in a few paragraphs, as shown in the examples on pages 25 and 27. Some longer, more detailed trip reports may need headings to break up the narrative into visible compartments. Typical headings might be:

Summary

Assignment Details:
    Purpose
    Authority
    Personnel   *(Background)*
    Location
    Duration of Trip

Assigned Work Completed

Problems Encountered   *(Trip Activities)*

Additional Work Done

Results Achieved

Follow-up Action Required   *(Outcome)*

Trip reports are often written as interoffice memorandums from the person who made the trip (or was in charge of a group of people who did the job as a team) to that person's supervisor or manager. Often, too, they are written in the first person—"I" if the writer was alone, and "we" if several people were involved. The first person has been used in both example trip reports that follow. They comprise

- a report on a field installation (page 25), and
- a report on a site evaluation (page 27).

A method for adapting trip report compartments so they can be used to report conference attendance is suggested on page 26.

Trip Report No. 1
Reporting an Installation
Comments

Frank Crane is a field service representative for Vancourt Business Systems Inc. He is reporting an installation he has just completed to his company's R&D Manager, Dale Rogerson.

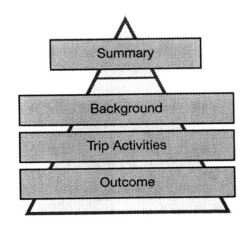

① For his **Summary Statement** (opposite) Frank has picked information primarily from the Background and Trip Activities compartments.

② The **Background** compartment brings together all the bits and pieces of information relating to the trip. (It answers the questions: *Who* went *where, why,* and *when?*) Note that Frank has spent some time describing the purpose of the modification kit, so that readers not familiar with it will be better able to understand his report.

③ The **Trip Activities** compartment starts here. For a trip report describing installation, modification, or repair work, the following guidelines provide a useful rule of thumb:
   • Describe routine work that goes according to plan as briefly as possible, particularly if there is an instruction or work specification that can be referred to and attached.
④ • Describe unexpected work, unusual events, or problems in some detail, and particularly explain how a difficult situation has been resolved.

⑤ *Note:* To conserve space in this book, the diagram and invoice have been omitted.

⑥ Sometimes it can be difficult to identify exactly where the **Outcome** compartment starts. Some people might say it starts at the beginning of the previous paragraph; others might say it starts at this paragraph; while still others might argue it starts at the next paragraph. Only Frank really needs to know where the Outcome starts (because he uses it to organize his report). All that is necessary from the reader's point of view is that the report reads smoothly and progresses logically from beginning to end.

⑦ In this final **Outcome** statement, Frank provides a "memory jogger" for Dale Rogerson (to whom the report is addressed) and Gerry Morganski (who receives a copy). In effect he is saying: *Take note! Plan to send someone to Westland between January 20 and 24.*

⑧ If you send a printed copy, a signature helps confirm and personalize the message. If you email the report, a signature isn't possible.

Vancourt Business Systems Inc.

TO: Dale Rogerson, Manager   **DATE:**   October 23, 2002
          Research and Development

FROM: Frank Crane, Field Service Rep.   **SUBJECT:** Installation of Prototype
                                                         Modification Kit MCR-1

(1) An MCR-1 multi-account readout display and control box have been installed on a model 261 Processor, where they will be field-tested for three months.

(2) I was assigned by Work Order M97 to install the prototype kit on a processor owned by Arrow Industries at Westland, Ohio, where arrangements have been made for it to be field-evaluated. Modification kit MCR-1 permits raw data on individual accounts stored in Vancourt 261 Processors to be made instantly available on a miniature display unit mounted beside the processor. I drove to Westland on October 19 and returned on October 22.

(3) The circuit and control box were installed without difficulty. However, a locally manu-factured equipment rack on which the 261 Processor has been mounted prevented me from installing the miniature display beside the processor, as directed in step 29 of the installation specification.

(4)
(5) I arranged for the mounting tray to be modified by Corwin Metals in Westland, so that it could be mounted on top of the processor as shown on the attached diagram. Corwin Metals's invoice for $246.25 is attached.

(6) During post-installation tests I detected and corrected two minor display faults. I then tested the installation operationally for three hours, but detected no further faults. During this period I trained three employees of Arrow Industries to use the equipment.

(7) Their contact employee is Lara Carter (one of the three persons I trained), with whom I have left evaluation and serviceability status report forms. She will mail these to you weekly.

(8) Field Services will have to arrange for a technician to return to Westland between January 20 and 24, 2003, to disconnect and remove the modification kit.

Frank

c: Gerry Morganski, Field Service Manager

Trip Report No. 2
Reporting a Site Evaluation
Comments

Tory LeValle is responsible for organizing a professional conference and is evaluating a potential hotel in a report (opposite) to Joshua Swanson, the society's president.

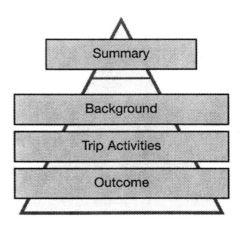

(1) This subject line is effective because it contains a word that describes what Tory is *doing*. She writes "Evaluation of..." (the hotel); if she had written "Visit to..." it would not have been as informative.

(2) Tory writes a positive **Summary Statement** that summarizes the result of her visit.

(3) This **Background** compartment does two things:
- It outlines Tory's involvement.
- It answers the questions *Who?, Why?, Where?,* and *When?*

(4) The next three paragraphs contain the **Trip Activities**. In this report, Tory explains what she saw and evaluates it against the requirements.

(5) For this report, the **Outcome** compartment is particularly important because it encourages or recommends action. It is written in a confident and strong tone.

Reporting Conference Attendance

The trip report compartments can be used to describe attendance at a conference or meeting. The most difficult one to write is the Trip Activities compartment, and the most efficient way to organize it is to divide it into subcompartments that focus on

- what you expected to gain, learn, or find out by attending the conference,

- what the program promised would be covered,

- what sessions you attended and why you chose them (this is important for a conference with several simultaneous sessions),

- what you gained or learned by attending these sessions,

- what you gained or learned from meeting and talking to other persons attending the conference, and

- what other activities you attended.

TO: Joshua Swanson   **FROM:** Tory LeValle
     President, Communication Society   2004 Conference Chair

(1) **SUBJECT:** Evaluation of Holidome as conference location

DATE: February 18, 2002

(2) I found that the meeting facilities at the downtown Holidome meet our requirements for holding the Communication Society's annual conference on October 12–16, 2004.

(3) As Conference Chair for the 2004 conference I am responsible for recommending a hotel that can accommodate our needs. On February 12, I met with Rene Phillips, the Facility Manager of the downtown Holidome, to discuss the feasibility and cost of holding our conference in their facility and to inspect the conference rooms, restaurants, and surrounding points of interest. While I was there, I also met James Jackson who is in charge of catering.

(4) The ballroom can comfortably seat 400 people, which is more than the maximum attendance we expect. Rene showed me 6 conference rooms that seat 40 people and 4 conference rooms that seat 20 people, all of which are available during our requested dates. I've confirmed these numbers with Elizabeth Marlow, our Program Chair, and she agrees that the rooms meet our requirements. The room charges will be $5250, which is $1250 more than we budgeted; however, because of the ideal downtown location, we expect we will attract local attendees to cover the difference.

The catering staff provided samples of their lunch options and a complete cost breakdown of their menu items. I'm confident we can select an interesting banquet within our budget of $45 per participant.

The Holidome provides a free shuttle service to all local attractions and restaurants, including the zoo, the art museum, and the Hard Rock Café. This is an added bonus for our attendees.

(5) Because of the location, costs, and facilities, I suggest we sign a formal contract with the Holidome to guarantee our required dates.

Tory

Inspection Reports

An inspection report is similar to a field trip report in that its writer has usually gone somewhere to inspect something. Bev Hubka's trip to Jackson, Mississippi, to inspect damaged computers is an example (see Chapter 2).

Other typical situations that would require you to write an inspection report include

- examination of a building to determine its suitability as a storage facility,
- inspection of construction work, such as a bridge, building, or road,
- checks of manufactured items, to assure they are of the required quality, and
- inspection of goods ordered for a job, to check that the correct items and quantities have been received.

The writing compartments are also similar to those for a trip report, except that the compartment previously labeled "Trip Activities" is more clearly defined as Findings. The report writing pyramid is illustrated in Figure 3–3.

The following notes suggest what these compartments should contain and how the information should be arranged.

Figure 3–3   Writing compartments for an inspection report.

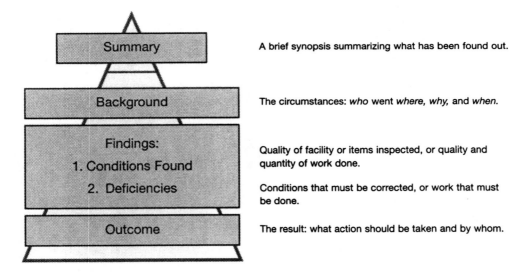

- The **Background** compartment describes the purpose of the inspection, mentions on whose authority it was performed, and lists circumstantial details such as the names of people involved and the date and location of the inspection.

- The **Findings** compartment is best divided into two subcompartments, one to describe *conditions found* and the other to list *deficiencies* (a deficiency can be either an unacceptable condition or a missing item). The length and complexity of the findings dictate how these compartments are organized. Short, simple findings are arranged in this order:

Conditions Found:
    1. _____
    2. _____
    3. _____ (etc)
Deficiencies:
    1. _____
    2. _____ (etc)

The arrangement is illustrated in Inspection Report No. 1 [see items ④ and ⑤ on page 33].

Longer, more complex findings should be arranged so that the deficiencies for each item are listed immediately after describing the item's condition:

Inspection Findings:
    Item A:
        Condition
        Deficiencies
    Item B:
        Condition
        Deficiencies (etc)

This arrangement is shown in Inspection Report No. 2 [see items ⑥ to ⑪ on pages 35 and 36]. The intention is to keep the deficiencies reasonably close to the condition from which they evolve.

Try to use a miniature pyramid to organize the information in each of the Conditions Found and Deficiencies subcompartments, as shown in Figure 3–4.

Figure 3–4   Organization of the *Conditions Found* and *Deficiencies* subcompartments.

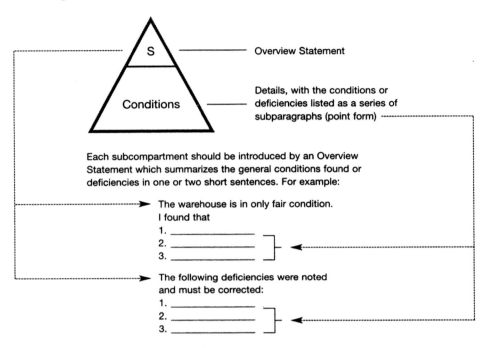

- The **Outcome** compartment suggests what should be done as a result of the inspection. If deficiencies have been listed at the end of the previous compartment, the outcome is likely to be short. For example:

Providing the deficiencies I have listed are corrected, the warehouse should make a suitable storage facility for the Passant Project.

The two inspection reports on the following pages demonstrate how these guidelines are applied. Notice how the first person ("I") is used by both writers. It creates a strong, confident image and avoids writing in the passive voice.

A form for recording inspection information is shown in Figure 3–5. The person carrying out an inspection makes brief notes of the conditions and deficiencies directly onto the form while on site and then later transcribes the information into a written report. Alternatively, and particularly for short reports, the form can be used as the final reporting document as long as the notes are complete thoughts that the intended reader will understand.

Figure 3–5   A form for an inspection report.

INSPECTION REPORT

Location: _____   Date: _____

Item(s) being inspected: _____

Inspector: _____   Contractor: _____

Conditions Found:

Deficiencies:

Recommendation(s):

Inspection Report No. 1
Inspecting a Contractor's Work
Comments

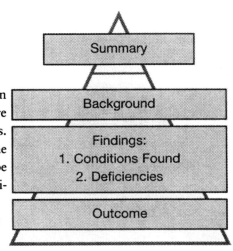

(1) Paul Thorvaldson, the author of the report on the opposite page, has chosen a slightly more formal format than a memorandum offers. He reasons that several people will read the report, and the contractor will probably be given a copy when asked to rectify the deficiencies.

(2) The **Summary Statement** tells readers right away what they most want to know—that the new facility is ready.

(3) In the **Background** compartment, Paul describes the details leading up to his inspection visit. He answers the questions *Who?, Where?, When?,* and *Why?*

(4) The **Findings** compartment starts here with an overview statement ("The contractor has done a good job."). It continues immediately with the **Conditions Found** subcompartment, which Paul limits to two short sentences mentioning the main items he noticed. When specifications have been met, there is no need to describe everything that has been done; it is sufficient simply to indicate that the job has been completed correctly. But when specifications have not been met, attention must be drawn to every item that has been improperly done [see (5)].

(5) The **Findings** compartment continues with the **Deficiencies**. Each item needing correction is listed in a separate subparagraph (to make it easy to identify step by step what action has to be taken) in clear-cut terms that will not be misunderstood. If there are many deficiencies it may be more convenient to list them on a separate sheet or sheets (called an attachment), and to refer to them in the Deficiencies paragraph:

> The 27 deficiencies listed on the attached sheets must be corrected by the contractor before the job can be accepted.

Paul numbered the deficiencies so they can easily be referenced.

(6) The **Outcome** compartment describes the result of the inspection (in this case, whether the contractor's work can be accepted and the new accommodation occupied). The Outcome compartment provided Paul with the primary information he needed for the summary at the start of his report.

Robertson Engineering Services

INSPECTION REPORT

**(1) Alterations to and Redecoration of New Offices
for the Technical Publications Department**

(2) Except for some minor deficiencies, the repair and renovation work is complete. The offices can be occupied on November 1, as scheduled.

(3) These offices were previously occupied by Nor-West Distributors, who vacated them on September 30, 2001. A contract for renovating and redecorating the premises was let to Craven Builders Inc on October 2; it specified the work to be done and a completion date of October 30, 2001. The contractor notified us on October 26 that the work was complete and I inspected the premises on October 27.

(4) The contractor has done a generally good job. There are no signs that the previous temporary walls existed or have been removed, and the new temporary walls look like permanent structures. Decorating quality is good.

(5) I noticed the following deficiencies, which the contractor must correct.

1. The rubber underlay needs to be relaid under parts of the fitted carpet in the northeast corner of the main office. Currently it is bunched in seven or eight places.

2. The pelmets over all four windows need to be extended to the specified 78-inch width. Currently they are only 64 inches wide.

3. The door to the manager's office needs to be rehung so that it closes fully and the lock engages.

4. Four of the lighting fixtures need to be realigned so that all are at the same level.

5. The cove needs to be installed at the foot of all walls.

(6) These deficiencies will not prevent the Technical Publications Department from occupying the offices on November 1, although I suggest that furniture should not be placed in the northeast section of the main office until the carpet underlay has been relaid. All other deficiencies can be corrected without interfering with the department's operations.

I suggest that a second inspection be scheduled before contractor payment is approved.

Paul Thorvaldson

Paul Thorvaldson, PE

October 27, 2001

Inspection Report No. 2
Inspecting Electronic Equipment
Comments

Maxine Griffin has been inspecting computer equipment at a small consulting firm whose owner is about to retire and plans to close down the business. The company Maxine works for (Tri-State Engineering Services) has been given the opportunity to purchase the electronic equipment. She reports her findings (opposite) to Tri-State's Vice President, Wendell McKnight.

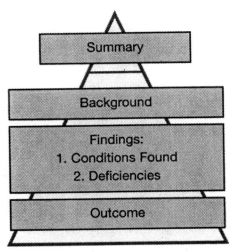

(1) This is Maxine's **Summary Statement,** in which she answers Wendell's immediate question: "Should we buy the computers and equipment?"

(2) In the **Background** compartment Maxine answers the questions *Who?, Where?, When?,* and *Why?* She gives more details than Wendell probably needs because she is aware that he may circulate her report to other managers who will be involved in the purchase decision, and they may not know the circumstances.

(3) Her **Findings** compartment starts with a two-sentence overview statement that prepares readers for the specific details that follow. By naming the equipment groups (Palmtop computers, digital cameras, All-in-One printer, LCD projector) in the order in which she will discuss them, she subtly introduces her readers to the report's overall plan.

(4) The attachment mentioned in the report has been omitted to conserve space in this book.

(5) By using headings, Maxine introduces variety into what would otherwise have been a long, dull-appearing narrative.

(6) This is the first **Conditions Found** compartment. A separate Conditions Found paragraph follows each of the three next headings.

(7) This is a combined evaluation and **Deficiencies** compartment for the first set of equipment. Maxine has chosen to use the alternative arrangement, described on page 29, and so presents the evaluation and names the deficiencies immediately after describing the Conditions Found for each equipment.

(8) The first paragraph after the "Digital Cameras" heading is Conditions Found.

Tri-State Engineering Services

To: Wendell McKnight

Subject: Inspection of Hawthorne and
Associates' Electronic Equipment

From: Maxine Griffin *mg*

Date: February 18, 2002

(1) I have inspected Hawthorne and Associates' computer equipment and consider that approximately 83% is worth purchasing for a total cost of $4300. This equipment is three years old or less and would cost at least $7000 if we were to purchase it new, even at today's discounted prices.

(2) My visit to Hawthorne and Associates was scheduled following Sam Hawthorne's February 14 call to you inviting us to buy the equipment before he offers it for public sale on March 1, the day after his business ceases operation.

(3)(4) The equipment comprises two Palmtop computers, two digital cameras, an All-in-One printer, and an LCD projector. It was purchased new between 1999 and 2001 and is itemized in Attachment 1.

(5) **Palmtop Computers**

There are two Palmtop computers, both Nabuchi model 1700 XL running the entire MS Office software. Each has a 8 GB (gigabyte) hard drive, a 56K internal modem, and 120 MB RAM, which is sufficient to run our presentation software. They were purchased in 2000 for $1675 each. Mr. Hawthorne has priced them at $800 each, which includes the computer, battery pack, external CD player, and a padded carrying case for each. The extended warranty agreements can be transferred to us: they are valid until October 2004.

(6)

(7) Both computers are fully operational but are showing wear after nearly two years of use. One has a slight crack in the screen casing, which will invalidate the extended warranty on the active matrix screen. The processor and small hard drive will limit their capability in the future. They will not be able to be used with our current network software since the hardware is outdated and cannot be configured for the new software protocols. Additionally, the display screens of both are washed-out and probably have only a short life remaining.

The small keyboard will take some getting used to, but if the computer is used only for email and presentations, it shouldn't be a major problem. These computers could be used by executives who travel frequently and need to stay on top of their email and present data during meetings with our clients.

Digital Cameras

(8) The digital cameras are Sony ENCIA ED 81s and come with 1024 x 768 resolution and a 3X zoom capability. The built-in microphone and speaker allow for recording and playback of sound memos. The original price in March 2000 was $799 each. Mr. Hawthorne says we can have them both for a total of $900.

(9) Although the technology has improved, the features and capabilities of this particular model will suffice for how we will be using them. Since we've never used digital cameras before, these features seem to be attractive and adequate. These cameras will be valuable to our engineers during their site inspections and to our Marketing Department when developing the online product catalogue.

All-in-One Printer

(10) The Jansen 405 All-in-One printer allows color printing and copying, and electronic distribution of hard-copy documents to fax and email addresses. It can also receive faxes and print full-sized posters. It would be an appropriate replacement for the outdated Hewlett-Packard LaserJet III in the Marketing Department. The original purchase price was $2500 and Mr. Hawthorne has priced it at $1800.

(11) The All-in-One printer is in excellent condition. I don't think the Hawthorne employees used it that much. It was originally purchased in 1999 and they have consistently maintained it with high-quality toner and parts. I have researched the Jansen website and this particular model is still very popular and provided with industry software upgrades.

LCD Projector

(12) The LightBox 705 LCD projector comes with a hard-cover carrying case with wheels. The size and weight are at the maximum carry-on allowance according to airline guidelines. When it was purchased in 1999 it was the best available. The internal light source is adequate so the lights in a conference room can remain on. The speaker system is not working and this is a drawback since our sales presentations usually include clips from our marketing videos. An extra set of speakers would need to be purchased. In 1999 the price was $8750 and Mr. Hawthorne is selling it for $6800.

(13) The major improvements in today's models are in size and weight. This projector weighs twice as much as a new one would. Although we need an LCD projector, I think we should research the current market and compare prices. Our salespeople already complain about the amount of equipment they have to carry to customer visits. This additional piece would add considerable weight. I agree we need an LCD projector of our own since relying on our clients' equipment often leaves us with compatibility issues.

Recommendation

(14) I recommend we purchase the two Palmtop computers, the two digital cameras, and the All-in-One printer for a total of $3400 from Hawthorne and Associates. Since newer LCD projectors have come down in size and cost, I also recommend we investigate how much a new LCD projector would cost before we consider purchasing Hawthorne's older model.

(9) This is the second evaluation/deficiencies compartment.

(10) This paragraph describes Conditions Found.

(11) There are no deficiencies for this attractive item, so Maxine emphasizes the positive qualities.

(12) Here is another paragraph of **Conditions Found.**

(13) This is the evaluation/deficiencies of findings in the previous paragraph [at item (12)].

(14) The **Outcome** starts here. Maxine draws a conclusion and then presents an alternative for her readers to consider. It was from this paragraph that she drew most of the information for her Summary Statement [at (1)]. The **Outcome** compartment continues with a firmly worded recommendation. Maxine's use of the first person ("I") prevents her from inadvertently writing a weak recommendation such as "It is recommended that ... "

CHAPTER 4

Progress Reports, Project Completion Reports, and Short Investigation Reports

Like the reports in Chapter 3, these informal reports are also short, seldom exceeding three pages plus attachments. Their writing compartments, however, are often expanded to include more subdivisions, particularly in the Facts & Events compartment of the basic pyramid (Figure 2–3).

When a report contains detailed information, such as lists of materials, cost analyses, schedules, or drawings, these documents are normally removed from the body of the report (from the Facts & Events compartment) and placed at the back, where they are referred to as "Attachments" or "Appendixes." (This is done to avoid cluttering the report narrative with tabular data and thus interrupting reading continuity.) Because they provide supportive evidence, or "backup," for statements made in the report, a separate compartment is created for them at the foot of the report writer's pyramid (see Figure 4–1). This compartment is labeled **Evidence** and is often shown with a dotted line to indicate that it is optional.

Progress Reports

Progress reports keep management informed of work progress on projects that span a lengthy period, which can vary from a few weeks for a small manufacturing contract to several years for construction of a hydroelectric power station and transmission system.

Figure 4–1   The report writer's pyramid with an evidence compartment for attachments and appendices.

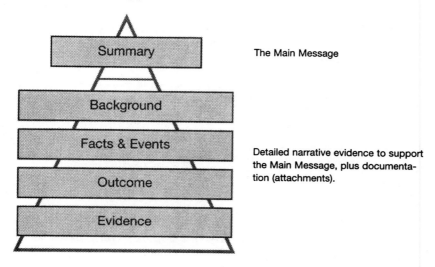

There are two types of progress reports:

1. Occasional progress reports are written at random intervals and usually concern shorter-length projects.

2. Periodic progress reports are written at regular intervals (usually weekly, biweekly, or monthly) and concern projects spanning several months or years.

The writing compartments are the same for both reports, although there are differences in their application. They evolve from the basic report writing pyramid (see Chapter 2), with two of the compartments relabeled to suit a progress-reporting situation (see Figure 4–2).

• **Progress** replaces the basic Facts & Events compartment and is subdivided into four smaller compartments, which describe
    1. planned work,
    2. work done,
    3. problems encountered, and
    4. adherence to schedule.

• **Plans** replaces the original Outcome compartment. There is also an optional **Evidence** compartment, for assembling forms and statistical data pertinent to the project.

These compartments are described in more detail on the following pages, with references to the two progress reports between pages 43 and 48.

Figure 4–2   Writing compartments for a progress report.

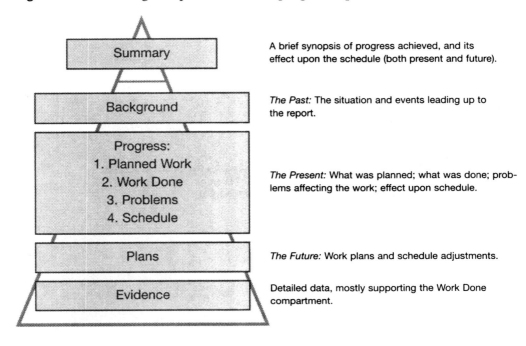

A brief synopsis of progress achieved, and its effect upon the schedule (both present and future).

The Past: The situation and events leading up to the report.

The Present: What was planned; what was done; problems affecting the work; effect upon schedule.

The Future: Work plans and schedule adjustments.

Detailed data, mostly supporting the Work Done compartment.

Occasional Progress Report

Occasional progress reports apply to short projects during which probably only one progress report will be necessary. Sometimes they are written near the project's midpoint. Occasionally they are written to forewarn management that problems have occurred and delays can be expected. But most often they are written as soon as the project leader has a sufficiently clear picture of work progress to confidently predict a firm project completion date.

The report writing compartments for an occasional progress report are shown in Figure 4–2 and are discussed here in more detail.

- The **Summary Statement** should comment briefly on the progress achieved and whether the project is on schedule; it may also predict a project completion date. Its information is drawn from the Work Done, Schedule, and Plans compartments.

- The **Background** compartment should describe briefly the people involved in the project, and the location and dates (i.e. it should answer the questions *Who?*, *Where?*, *Why?*, and *When?*). If only people familiar with the project will read the report, then only minimum background information is necessary.

- The **Progress** compartment contains information from the four subcompartments illustrated in Figure 4–2, which are normally arranged in the order shown (although it is not uncommon for some of these subcompartments to overlap or be omitted).

  1. The **Planned Work** subcompartment outlines what work should have been completed by the reporting date. Normally only a brief statement, it can refer to an attached schedule or work plan.

  2. The **Work Done** subcompartment describes how much work has been completed. Only brief comments are necessary for work that has gone smoothly and has progressed as planned. If lengthy numerical data has to be included, it should be placed in an attachment rather than inserted in the report narrative. More detailed comments should be provided if there have been variances from the work plan. They should explain why the variances were necessary and any unusual action that was taken.

  3. The **Problems** subcompartment comprises events or situations that affected the *doing* of the job (e.g. a blizzard that stopped work for two days, late delivery of essential parts, or a strike that prevented access to necessary data). These problems should be described in detail, and the explanation should include what action was taken to overcome each problem and how successful the action was.

  4. The **Schedule** subcompartment states whether the project is ahead of, on, or behind schedule. If ahead of or behind schedule, the difference should be quoted in hours, days, or weeks.

- The **Plans** compartment is usually short and describes the report writer's plans and expectations for the remainder of the project. It should indicate whether the project will finish on schedule and, if not, predict a revised completion date. There should be an obvious link between this compartment and the previous subcompartment (Schedule).

- The optional **Evidence** compartment contains data such as drawings, statistics, specifications, and results of tests, which if included with the earlier parts of the report would clutter the report narrative. This supporting information is grouped and placed in attachments. Each attachment must be referred to in the Background or Progress section of the report, so that the reader will know it is there.

These compartments are identified in Progress Report No. 1, on the following two pages.

Progress Report No. 1
Occasional Progress Report
Comments

Marjorie Franckel is studying the effect that a proposed high-voltage power transmission line through Alaska and the Yukon will have on the traditional calving grounds of the Porcupine Herd of caribou. Her environmental studies have been delayed and she is reporting her progress to Vic Braun, her manager.

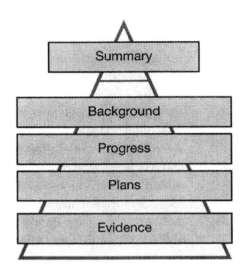

(1) This is the **Summary Statement**. It reports mainly that Marjorie's study is running behind schedule.

(2) The **Background** compartment also includes the **Planned Work** subcompartment. Although Marjorie is aware that Vic Braun knows these details, she includes them because she expects he will send a copy to the Department of Transportation.

(3) The **Progress** compartment starts here and continues to the middle of the paragraph labeled (6). The **Work Done** subcompartment also starts here. Quoting facts and figures, as Marjorie has done in the three subparagraphs, makes a report writer sound confident and knowledgeable.

(4) The attached map becomes the **Evidence** compartment. (It has been omitted from this copy of the report to conserve space.)

(5) These are more of Marjorie's findings and so are part of **Work Done**.

(6) In this **Problems** subcompartment, Marjorie outlines why part of her report is rather vague and why the study has been delayed.

(7) The **Schedule** subcompartment is this single, rather indefinite sentence. Marjorie cannot be more exact because she simply does not know how long it will take to find and interview people.

(8) This final paragraph is her **Plans** compartment.

H L Winman and Associates

Interoffice Memorandum

TO: Vic Braun, Manager
Environmental Studies

FROM: Marjorie Franckel
Biologist

SUBJECT: Progress: Study of
Caribou Calving Grounds

DATE: August 18, 2002

(1) My study of the caribou calving areas used by the Porcupine Herd of caribou has been delayed by lack and inaccessibility of data. I doubt whether I will be able to complete the study before September 15.

(2) The study is being done for the Northern Power Commission to determine the boundaries and dates of calving so that specific areas can be designated as "Restricted Zones" during the calving season. Currently I am working out of Old Crow in the Yukon.

(3) I have defined the eastern and western limits of the North Slope calving area bordering the Beaufort Sea (see attached map), and have identified three approach routes used by
(4) the caribou during their northbound spring migration. These are:

- Through the Richardson Mountains in the east, along the Yukon/Northwest Territories border.

- Through the Brooks Range of mountains north of Old Crow.

- Through the Brooks Range in Alaska, between the Canning River and the Yukon border.

(5) In normal years most calving seems to take place in the Arctic Wildlife Refuge in Alaska between early May and early June. But if bad travel conditions delay the migration, calving occurs farther east along the central plain or sometimes even in the mountain ranges while the herd is still migrating.

(6) My problem has been to identify which migration routes are most used, clear-cut dates when calving occurs, and the earliest and latest dates that the caribou have been known to enter the coastal plain. Only a few residents have observed calving, and I have been trying
(7) to identify who they are and to interview them. This lack of real information has delayed my study by at least 15 days.

(8) For the next two or three weeks I will be travelling with an interpreter to interview Inuit in very small communities north of Old Crow and as far east as Aklavik. During this period it is unlikely you will be able to contact me.

Marjorie

Periodic Progress Report

The compartments for a periodic progress report contain similar information to those for an occasional progress report, but there is some shift in content and emphasis.

The format of a periodic progress report also appears to be more rigid than that of an occasional report. This rigidity is imposed not so much by established rules as by the content and shape of the initial report in a series. The implication is important: report writers should take great care in planning a progress report that is to be the first of a series because they will be expected to conform to the same shape in successive reports.

The compartments outlined below are those shown in Figure 4-2. They provide useful guidelines and they demonstrate the differences in content between the occasional and periodic progress reports.

- The **Summary** should comment briefly on the work accomplished during the reporting period. It may also mention whether the project is on schedule. This information can be drawn from the Work Done and Schedule compartments.

- Except for the first report in a series, which will be fairly detailed, the **Background** compartment probably will refer only to
  - the project number or identification code,
  - the dates encompassing the specific reporting period, and
  - the situation at the end of the previous reporting period, with particular reference to the project's position relative to the established schedule.

- The **Progress** compartment is divided into four subcompartments. In short reports these subcompartments may interlock or overlap, but in longer reports they are more likely to be independent units. If there is no information for a particular compartment, then the compartment is omitted.
  1. The **Planned Work** subcompartment outlines what should have been accomplished during the reporting period. It may refer to either the original schedule or a revised schedule defined in a previous progress report. Normally it is short, sometimes it is combined with Work Done, and occasionally it can be omitted.
  2. The **Work Done** subcompartment describes what has been achieved during the reporting period. Ideally, this subcompartment will
     - open with a brief overview statement that sums up in general terms what has been accomplished,
     - continue with a series of subparagraphs each describing in more detail what has been done on a specific aspect of the project,
     - refer to attachments containing comprehensive numerical data, statistics, or tables (see the Evidence compartment), and
     - explain variations from the planned work, or unusual activities affecting work progress (this may be linked with the Problems subcompartment).

3. **Problems** are factors that have caused changes in plans or in the schedule. The report should describe what action has been taken to overcome the problems, whether the action was successful, if the problems still exist, and what action will continue to be taken, either to avert the problems or to make up lost project time.

4. The **Schedule** subcompartment states whether the project was ahead of, on, or behind schedule on the last day of the reporting period. (There may be a convenient link between this compartment and the end of the previous compartment.) If ahead of or behind schedule, it should state the number of hours, days, or weeks involved. It may also predict when the project will be back on schedule and recommend a revised schedule for the next reporting period.

- The **Plans** compartment is very short if the project is running smoothly and is on schedule. But if there are problems affecting the work, it should outline the report writer's expectations for the next reporting period or even suggest a revised schedule for the whole project.

- The optional **Evidence** compartment is used to store detailed information such as forms containing weekly summaries of work done, tests, and inspections.

Progress Report No. 2
Periodic Progress Report
Comments

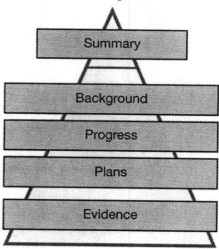

① Project coordinator Roger Korolick has chosen this semiformal format, rather than a memorandum, for his progress report on a trailer installation (page 46) because it will be circulated to several people, such as the Divisional Vice President and the Marketing Manager.

② This subheading takes the place of the **Background** compartment.

③ Roger has taken main points from the Work Done compartment for his summary. ("Synopsis" is an alternative but less used name for "summary.")

④ The four main headings and the paragraph numbering system help readers *see* how Roger has organized his report.

⑤ This overview statement at the start of the **Progress** compartment gives the reader a quick picture of the overall situation and introduces the details that follow.

⑥ Roger's description of **Work Done** starts here and continues to the end of ⑦. He reports on the exterior painting first because it has been carried forward from the

Vancourt Business Systems Inc.

① Progress Report No. 5 – Project W16

② Equipping Electronic Equipment and Display Trailer
Reporting Period: June 1 to 30, 2002

1. Synopsis

③ The project is generally on schedule. Trailer painting is complete, air conditioning has been installed, and work has started on wiring and positioning the work stations.

2. Work Accomplished

④ ⑤ With the exception of the instructor's console, work has progressed well during the month. Major activity has occurred in five areas:

⑥ 2.1   **Exterior Painting.** The 30-ft Fruehauf trailer, which had been taken to Display Signs Inc. on May 26, was returned on June 8 with the corporate logo and trailer identification painted on the back and sides.

⑦ 2.2   **Individual Learning Centers.** The first three of the six carrels being built by Carpenters Unlimited were received on June 15. Our electrician has wired up two of them and they have been installed on the left-hand wall of the trailer. This component of the project is on schedule.

2.3   **Instructor's Console.** Snags have again interrupted construction. Frank Dartmouth, in custom manufacturing, has identified the main problem as late delivery of modules from Capricorn Electronics in Toledo, which has set his assembly schedule back by 15 workdays. He now predicts a completion date of August 19.

2.4   **Display Booths.** The two booths for displaying the prototype electronic equipment are complete and ready for installation in July. Marketing has assembled and tested the display equipment, and I have arranged for the two systems to be installed during the first two weeks of August.

2.5   **Air Conditioning.** The air conditioning unit was installed by the Kool-Air Company between June 12 and 16. Initial use of the air conditioner has identified two unsatisfactory conditions:

   (a)   Moisture is running back into the trailer and dripping onto the newly installed carrels.

⑧ (b)   There is severe vibration and noise from the air conditioner.

⑨ The installation contractor has examined the air conditioner and will return on July 5 to correct these conditions.

previous month (his May 31 report mentioned that the trailer had been taken out for painting).

(7) The remaining four items all describe work that will continue into the following months. For each, Roger finishes with a closing statement that comments on the schedule, predicts a revised completion date, or states what action will be taken and when. This is simpler than carrying all the plans information forward and lumping it into one big compartment later in the report.

(8) The second of these "unsatisfactory conditions" introduces a problem, which Roger develops further in paragraph 3.1.

(9) Roger must remember to carry this item forward and comment on it in the Progress section of his July report.

(10) This is the **Problems** subcompartment. In paragraph 3.1 Roger forewarns management of a situation that may develop into a serious problem. He must comment on it again in next month's report, as he has done in paragraph 3.2 for the previous month's problem.

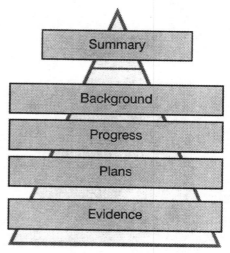

Normally, a previous month's problems are discussed before new problems are introduced (i.e. the continuing short-circuit problem in paragraph 3.2 should be mentioned before the new noise problem in paragraph 3.1). Roger has chosen to discuss the noise problem first because he wants to maintain continuity from the description of the unsatisfactory condition in the previous paragraph (paragraph 2.5).

(11) Roger combines the **Schedule** subcompartment and **Plans** compartment under one heading because the two topics are interrelated. Note how his closing statements in paragraphs 2.3 and 2.4 prepare readers for the changes in plans he announces in paragraph 4.1.

(12) The attachment mentioned here forms the **Evidence** compartment. (It has been omitted from the report to conserve space.)

3.   **Problems Encountered**

3.1   I am concerned that the air conditioning unit will create a noise problem that will be unacceptable for the training we are planning. Already, installation personnel working in the trailer have commented that the noise is unusually high. Although the contractor has assured me that the noise level will drop significantly when the vibration problem is corrected and the carrels, booths, and carpet are installed, I doubt that the drop will be sufficient. We may have to find a means for further lessening the noise.

3.2   The intermittent short-circuit condition in the trailer strip lighting, described in para 3.3 of my May report, was traced to a faulty switch and corrected on June 6.

4.   **Scheduling**

4.1   To overcome the problem created by late delivery of the instructor's console, I plan to advance the installation of the display booths by three weeks (to start on August 1) and to delay installation of the instructor's console until August 22. These changes are shown on the revised workplan attached to this report.

4.2   I expect the project to be completed by September 15, as scheduled.

R. Korolick

Roger Korolick
Project Coordinator
July 2, 2002

Headings and Paragraph Numbering

Periodic progress reports and project completion reports can often benefit from the judicious use of headings and a simple paragraph numbering system. The following headings are suggested for a progress report.

Summary

Introduction

Project Progress

Problems Encountered

Adherence to Schedule:
    Current
    Predicted

Attachments

Use descriptive words or phrases as headings to help identify and explain the information in that section. A paragraph and subparagraph numbering system shows readers how the information has been organized and provides a handy means of cross-referencing within the report and between successive reports.

The suggested numbering system uses whole numbers for paragraphs and decimals for subparagraphs. For example:

3. *Project Progress*
    During the month we ...   *(Overview)*
    3.1 conducted field tests ...
    3.2 installed three pads ...   *(Specific Details)*
    3.3 removed old cables ...

We do *not* recommend continuing to a third level of decimals—i.e. 3.3.1, 3.3.2, etc—because the subparagraph numbers become too cumbersome. Instead, we suggest you use a single letter within parentheses: i.e. (a), (b), etc.

Periodic progress report No. 2 shows

- how writing compartments, headings, and a paragraph numbering system can be combined to shape a report, and

- how the writing compartments can be shaped to fit the requirements for each situation; they are not a rigid plan that you *must* follow.

Project Completion Report

A project completion report is equivalent to the very last of a series of progress reports. This is a very useful report that often does not get written because people are reassigned quickly to new projects. It may be a short one- to three-page report and should contain the project highlights and any unexpected conditions or work. It may also contain a compartment with suggestions of what to do if the project is to be repeated. In some industries this is what is called a "lessons learned" report.

Like a progress report, a project completion report has five writing compartments:

- The **Summary** identifies that the project is complete and states whether there is any special information the reader needs to know.

- The **Background** compartment states who authorized the project, its starting date and planned completion date, and who was involved in implementing it.

- The **Highlights** compartment describes the most important aspects of the project.

- The **Exceptions** compartment identifies any variances that occurred from the original project plan and, for each variance, discusses why it was necessary, how it affected the project, what action was taken to lessen its effect, and whether any further action is necessary.

  The final compartment can be

- an **Outcome** if the project is complete and no further work needs to be done, or

- an **Action Statement** if further work is necessary (in which case it identifies what needs to be done, and when and by whom it must be carried out).

Project Completion Report
Reporting a Project Is Finished
Comments

In a short one-page report, Joshua Lawrence explains he has finished the work requested of him and describes the additional work he completed because he discovered problems.

①Joshua is writing to the Engineering Quality Group's electronic media chairperson.

②In his **Summary** Joshua states what the project was and that it is finished. Joshua also briefly mentions what action is expected of the reader.

③In the **Background** Joshua states who authorized the project, what exactly he was supposed to do, and when the work was to be done.

④Joshua uses the **Highlights** compartment to describe the major accomplishments of the project and how he achieved them. Using specifics helps the reader understand the quantity of work.

RGI International

Specializing in Communicating Technical Information

September 30, 2001

(1) Dear Bethany Rabonivich:

(2) The testing and implementation of the Engineering Quality Group (EQG) website is now complete. I have uploaded all the files to the ISP server and indexed the site on 10 search engines. It will go live as soon as you have reviewed it and give me your approval.

(3) On September 10 the EQG president, Janice Sloboski, contacted me and asked me to test the links on their current website. The initial contract was to ensure all the links were valid and any referenced sites were still operative. I agreed to start the testing on September 14 and have it complete by September 26.

(4) I tested 74 webpages with 342 links. I also confirmed that the 16 referenced websites were still active and that the content was still related to the EQG's website information. Three referenced websites were no longer active so I researched the organization responsible for the information and found that two of the sites were combined to form a new site and the third site has fallen through the net. This means they no longer have a web presence. I updated the EQG site to reflect these changes.

(5) While I was testing the links I noticed that the home page did not have a usable navigational map. I performed some simple usability testing on the site, which showed most users became disoriented when using the EQG site and couldn't find the information they were looking for. Because of the potential frustration EQG members would experience when using the website, the president authorized me to redesign the home page, retest for usability, and extend the deadline by one week.

(6) I have completed all these tasks (the original testing, the redesign of the home page, and the usability testing) and the site is now ready for your review. You can access the information at *www.eqg.org*. When you have completed your review, please contact me so I can activate the site for the general public.

Sincerely,

Joshua Lawrence

Senior Consultant

⑤ As in most projects, Joshua discovered there was additional work that needed to be done that was not in the original scope of the project. He uses the **Exceptions** compartment to identify what he found out and what he did to correct it. He also explained the impact or effect the problem would have on the overall outcome of the project if left unresolved. Because the extra work changed the original delivery date, Joshua mentions who authorized the extra time in the schedule. He had previously submitted a Progress Report, which outlined the exact problem and the time and money implications.

⑥ Joshua's final compartment is an **Action** statement since he needs the reader to review the work and respond to him.

Depending on the size and length of a project, you could add an **Evidence** compartment at the end and attach all the previous Progress Reports to support the information in the Project Completion Report.

Short Investigation Reports

Most investigation reports are longer reports that examine a problem or situation, identify its cause, suggest corrective measures or ways to improve the situation, evaluate the feasibility of each, and select which is most suitable. These reports are discussed in Chapter 6. There are occasions, however, when only a minor or local problem is examined and only a short, informal investigation report is needed to describe it. Such reports are described here.

The short investigation report has the four basic compartments described in Chapter 2 plus the optional **Evidence** compartment. These compartments are illustrated in Figure 4–3 and outlined in more detail below.

Figure 4–3   Writing compartments for a short investigation report.

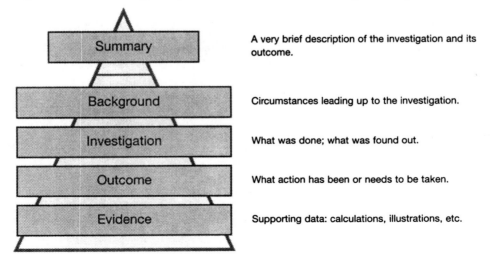

Summary	A very brief description of the investigation and its outcome.
Background	Circumstances leading up to the investigation.
Investigation	What was done; what was found out.
Outcome	What action has been or needs to be taken.
Evidence	Supporting data: calculations, illustrations, etc.

- A **Summary Statement** briefly identifies the problem and how it was or can be resolved.

- A **Background** compartment outlines what caused the investigation to be carried out.

- The **Investigation** compartment describes the steps taken to establish the cause of the problem and find a remedy.

- The **Outcome** compartment describes what has been done to resolve the problem or, if other people have to take the necessary action, recommends what should be done and sometimes who should do it.

- The optional **Evidence** compartment stores detailed supporting data evolving from the previous three compartments.

Very short investigation reports are usually issued as interoffice memorandums, or occasionally as emails or letters. An example is printed on page 54.

Short Investigation Report
Correcting an Electrical Problem
Comments

In this uncluttered one-page report, Tom Westholm places his information confidently into the report writing pyramid's five compartments.

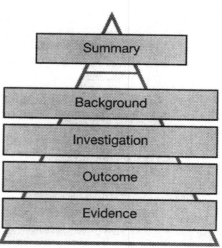

① Although the subject line is vague, the word "correcting" serves a useful purpose because it implies that Tom has found a solution to the problem.

② Tom's **Summary Statement** identifies the problem, states its cause, reports that it has been resolved, and suggests what else should be done.

③ In the **Background** compartment, Tom describes the events leading up to the investigation.

④ In his **Investigation** compartment, Tom describes his approach to the problem and what he has discovered.

⑤ Tom's **Outcome** compartment not only describes how he corrected the problem, but also suggests a better alternative.

⑥ Here, Tom refers to his **Evidence** information (which, to conserve space, has been omitted).

⑦ This final question is part of Tom's **Outcome** compartment.

Compare Tom Westholm's short, informal investigation report with Tod Phillips's five-page semiformal report in Chapter 6. Note how the five compartments used for the short report are expanded to develop more information for the longer report.

Interoffice Memorandum

To:	C. Meaghan, Plant Manager	Date:	July 9, 2001
From:	Tom Westholm Maintenance Electrician	Ref:	Correcting Electrical Blackouts

①

② I have traced the recent electrical power failures to a wiring error that created a power overload when two air conditioner compressors operated concurrently. Although I have corrected the problem, a better solution would be to install a separate power panel for two of the air conditioners.

③ The failures started after the air conditioners were overhauled in early May, and even then occurred only infrequently and at random intervals. On every occasion simply resetting the circuit breakers corrected the failure, which made the cause difficult to identify.

④ I compared the air conditioners' wiring connections against the manufacturer's wiring diagrams but could find no fault. I then physically examined the four conditioners individually and discovered a disconnected load splitter behind air conditioner No. 2. The load splitter was installed originally six years ago, to prevent the circuit from being overloaded should more than two air conditioner compressors cut in at the same time. Apparently the overhaul contractor failed to introduce the load splitter into the circuit when reinstalling the air conditioners in May.

⑤ I have reconnected the load splitter, but suggest we could obtain better performance from the air conditioners if we were to install a new power control box and ⑥ connect two of the air conditioners to it. We could then remove the load splitter. The cost would be $2245, as detailed on the attached cost estimate.

⑦ May I have your approval to buy the necessary parts and do the installation?

Tom

PART 3

Semiformal Reports and Proposals

CHAPTER 5

Test and Laboratory Reports

Considerable variation exists in the presentation of test and laboratory reports (often called lab reports). Some laboratory reports simply describe the tests performed and the results obtained, and comment briefly on what the results mean. Others are much more comprehensive: they open with a synopsis of the tests and results; they continue by presenting full details of the background, purpose, equipment, methods, and results; and they finish with an analysis from which their authors draw conclusions. The more comprehensive laboratory report is described here, because the shorter, simpler form can be adapted from it.

A third form of laboratory report is used in universities and colleges, where students are asked to perform tests and then write a lab report to describe their findings. Comments on laboratory reports written in an academic environment start on page 67.

Industrial Laboratory Reports

Industrial laboratory reports are based on the report writer's pyramid described in Chapter 2, but the **Facts** compartment is expanded to encompass the following four subcompartments:

1. Equipment & Materials

2. Test Method

3. Test Results

4. Analysis

Figure 5–1   Writing compartments for a test or laboratory report.

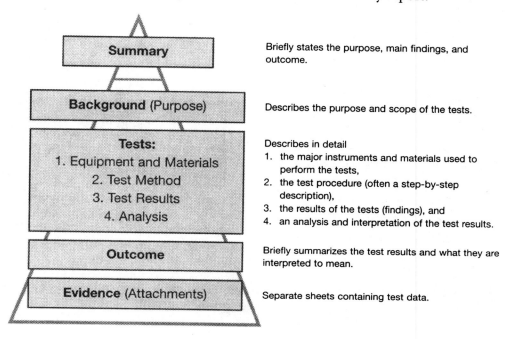

Summary — Briefly states the purpose, main findings, and outcome.

Background (Purpose) — Describes the purpose and scope of the tests.

Tests:
1. Equipment and Materials
2. Test Method
3. Test Results
4. Analysis

Describes in detail
1. the major instruments and materials used to perform the tests,
2. the test procedure (often a step-by-step description),
3. the results of the tests (findings), and
4. an analysis and interpretation of the test results.

Outcome — Briefly summarizes the test results and what they are interpreted to mean.

Evidence (Attachments) — Separate sheets containing test data.

There is also an **Evidence** compartment for holding specifications, procedures, and details of test measurements the author refers to. The major compartments and subcompartments are illustrated in Figure 5–1 and described in more detail on the pages facing the sample test report, which starts on page 59.

Industrial Laboratory Report
Testing a Water Stage
Manometer and Digital Recorder
Comments

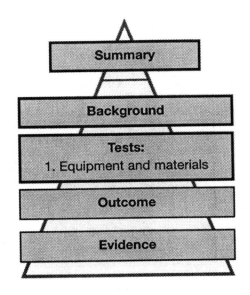

① The "title block" contains the report's title, number, and date. Some organizations use a prepared form for the first page of their laboratory reports. The form has spaces for entering predetermined information, such as
- title and purpose of test,
- name of client,
- authority for test (i.e. purchase order, letter, etc.),
- summary of test results,
- signature and typed name of person performing test,
- signature of manager or supervisor approving test results, and
- date tests were completed.

Report authors using a form find their Background compartment becomes much shorter or even nonexistent, because many background details are entered in the prepared spaces and need not be repeated.

② The **Summary** establishes what test was undertaken, sometimes why it was necessary, the main finding(s), and the result.

③ The **Background** compartment starts here. Carole Winterton (the laboratory technician who performed the tests and wrote the report) has chosen to divide the compartment into two parts, each with its own heading: *History* and *Purpose of Test*.

④ This is the start of both the large **Tests** compartment and the **Equipment and Materials** subcompartment. The amount of information provided under **Equipment and Materials** depends on several factors. A full description is provided if the client wants to replicate the tests or know more about how they were undertaken, or if the person performing the tests needs to demonstrate the extent of the testing. Only essential details are included if the client will be more interested in results and less concerned with how the tests were run. (The same guideline applies to the **Test Method** subcompartment.) If the equipment setup and list of materials are complex or lengthy, they can be placed in an attachment.

Environmental Test Labs Inc
Tests of Water Stage Manometer and Digital Recorder
from AWCS Site 24

Test Report No. 34/07

June 14, 2001

①

Summary of Test Results

② Tests of a solar powered Caledonia Water Stage Manometer model WSM, serial
No. 2306, and a Vancourt Digital Recorder model 2200, serial No. 10781, show
that all mechanical and electronic systems are operating satisfactorily. The
manometer, however, seems to have a minute internal gas leak.

History

③ The manometer and digital recorder were shipped to the Test Lab from site 24 of
the Agassiz Water Control System (AWCS). They were removed from service on
May 22, following a visiting technician's report that the digital recorder was indicat-
ing erratic changes in water stage that contradicted the technician's visual observation of water levels. The system could not be tested on site because site 24 is a
remote, unstaffed station with no power or physical facilities.

Purpose of Test

AWCS memorandum 0693 dated June 5, 2000, requested that Independent Test
Labs Inc. test the water stage manometer and digital recorder to determine
whether one of the instruments or the site's gas delivery system between the
manometer and the underwater orifice was causing the problem.

Equipment Setup

④ We installed the manometer on a workbench, levelled it 18° from the vertical, and
connected it to a mock-up of the site's gas delivery system, which consisted of

- a cylinder of super-dry nitrogen, through a gas flow register,
- 30 m of 5 mm ID polyethylene tubing terminating in a water tank, with the
  tube's orifice submerged 0.96 m below the surface (to simulate site conditions),
  and
- a Franck and Corwin type B-37 continuous recorder.

The digital recorder was connected to the manometer's output terminals, in parallel
with the B-37 recorder. The test hook-up is shown in Figure 1.

1

(5) A simple diagram showing how the test equipment is connected can help a reader visualize the test setup. If the illustration of a test setup is too large to fit on a standard page, it can be placed at the back of the report as an attachment and referred to at appropriate places within the report.

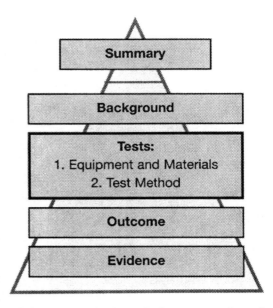

If several tests are performed, each with a varying arrangement of test equipment, it is better to insert a series of diagrams in the report, each positioned immediately ahead of or beside the appropriate test description.

(6) The **Test Method** subcompartment describes how the tests were carried out. It can range from a brief outline of the test method used (for a nontechnical reader interested primarily in results) to a step-by-step description of the procedure (for a reader who wants to know how comprehensive the test was). Carole has used a fairly detailed step-by-step description for her report, because her analysis of the results will depend on the reader fully understanding what was done during the tests.

(7) This initial paragraph is an internal summary statement that introduces a lengthy segment of the report. Preparing readers to expect a certain arrangement of information helps them accept more readily the facts a report writer presents. By numbering and naming the three tests, Carole is implying: "These are the tests you will read about next, and this is the sequence in which I will be presenting them to you." She must now take care to describe them in the same sequence.

(8) Each test is numbered and given a subheading similar to that used in the section summary statement (7).

(9) When a process or procedure is lengthy, or is not essential to a full understanding of the report, the steps may be printed in an attachment and simply referred to in the report. (See also comment (16).)

(10) Carole has chosen to place these five steps within the report, rather than in an attachment, because she wants readers to be fully aware that the first and fifth steps were performed *before* they read the test results and her analysis. They support her contention that the fault is unidentifiable and lies within the instrument she is testing.

In most instances, she introduces the steps in the active voice and then continues in the passive voice.

⑤

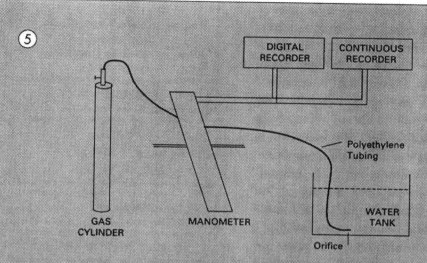

Fig. 1. Water Stage Manometer and Digital Recorder Test Setup

⑥ **Method**

⑦ We made independent tests of (1) the nitrogen delivery system, (2) the manometer, and (3) the digital recorder. We then tested the components as a system.

⑧ **1. Tests of Nitrogen Gas Delivery System**

We connected the gas flow regulator on top of the nitrogen cylinder to the polyethylene tubing (bypassing the recorder) and temporarily plugged the orifice in the water tank. Gas was allowed to flow into the delivery system until the regulator gauge read 500 kPa, and was stabilized at this level.
The tubing was inspected for leaks, but none were found, and the pressure remained stable for the ensuing two hours.

⑨ We reconnected the manometer to the system, removed the plug from the orifice, and applied nitrogen as specified in step 8 of the manufacturer's operating instructions (see attachment 1). When a flow rate of 15 bubbles per minute had been achieved in the manometer's sight feed, we checked the flow rate at the orifice. It was a steady 8 bubbles per minute, which conforms to the manufacturer's specification of one-half the flow rate at the sight feed, plus or minus 15%.

The sight feed and orifice bubble rates were checked at two-hour intervals during the tests and remained within specification throughout.

2. Check of Water Stage Manometer

We checked all external parts of the manometer for proper operation:

⑩
- The bubble flow rate in the sight feed was observed at two-hour intervals during the system checks, and at 15-minute intervals during the final four hours. It remained constant and within specifications.

- The float switch contacts were examined and found to be clean.

2

⑪ The **Test Method** subcompartment of a report must be written in clear, direct language which is entirely objective (unbiased, without opinions). This is one of the few occasions when the passive voice may be used in preference to the active voice. In the fourth bullet, for example, it would have been less comfortable for Carole Winterton to have asserted her presence by writing: "We removed and cleaned the constant-differential pressure regulator ... ." She also would have had to adopt the same construction for all the steps.

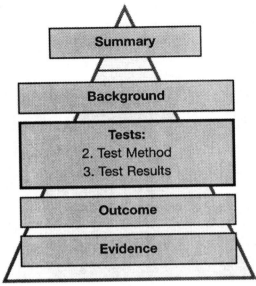

⑫ The pumping in and out of water has a direct bearing on, and is referenced in, the Results subcompartment. Consequently it must be described thoroughly here.

⑬ The **Test Results** subcompartment describes the major finding evolving from the tests. Like the tests described in the Test Method subcompartment, the test results must be written objectively.

⑭ In the **Analysis** subcompartment a report writer is expected to examine and interpret the test results and to comment on their implications. The analysis should discuss various aspects influencing or evolving from the tests and show how they lead to either a logical conclusion or an unanticipated outcome. This helps readers to understand and accept more readily the conclusions that follow.

- The set screw on the servo control was adjusted until the drive motor moved away from the correct setting. The servo control operated smoothly as it followed the adjustment.

⑪

- The constant-differential pressure regulator was removed and cleaned. No foreign matter was found.

- A soapy solution was applied to all exposed connections, both on the manometer and to the gas delivery system. No bubbling occurred.

3. Check of Digital Recorder

The digital recorder was inspected and recalibrated by Adanac Electronics Inc. No faults were found.

We then connected the recorder in parallel with the Franck and Corwin B-37 recorder and adjusted the manometer's servo control to simulate changes in water level. The readings on the digital and B-37 recorders tracked in parallel and indicated water levels within 0.025% of each other, which is well within the manufacturer's specification.

4. Check of Total System

We allowed the system to stabilize for 3 hours and then run continuously for 30 hours. For the first 26 hours the water in the tank was pumped in and out at controlled rates to simulate changes in water stage:

⑫

- Water was pumped out at 10.75 L/min (litres per minute) for 8 hours. The head of water above the orifice decreased from 0.96 m to 0.41 m.

- Water was then pumped in at 11.4 L/min for 12 hours, after which the water head had increased to 1.32 m.

- Water was again pumped out, this time at 8.4 L/min for approximately 6 hours, until the water head returned to 0.96 m above the orifice.

The system was run for a further 4 hours with no change in water level. Throughout, the manometer appeared to operate correctly.

Test Results

⑬

Tests of the gas delivery system and of the water stage manometer showed no apparent faults, but the digital recorder still showed that a fault existed somewhere in the system. Instead of showing a steady decrease, increase, and then decrease of water head, the recorder displayed a series of "steps," indicated by apparent abrupt changes in water level, each followed by a slow recovery (see attachment 2). These "false troughs" were present for increasing, decreasing, and stable water level conditions. When the chart on the Franck and Corwin B-37 recorder was inspected, it corroborated the digital recorder's output.

Analysis of Results

⑭

The erratic water level readings reported by the site, and the false troughs identified during the tests, are probably different interpretations of an identical fault.

3

⑮ This is the **Outcome** compartment. It should answer the question, resolve the problem, or respond to the request identified in the Background compartment (titled "Purpose of Test" in Carole's report). It must never introduce new data or present information that will surprise the reader. Carole does not make a recommendation, because her mandate was only to test the system and identify the cause of the problem.

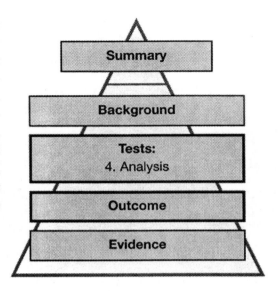

⑯ Attachments (page 66) form the **Evidence** compartment. Their purpose is to provide a place for storing data that a reader does not need while reading the report but may want to inspect later. They may comprise a detailed procedure used during the tests, a lengthy table of test results containing measurements and dial readings, or photographs, sketches, and drawings.

All attachments must be referred to in the report (the attachment on page 66 is referred to in the middle of page 2 of the report). They should be presented in the sequence in which they are mentioned in the report and then numbered sequentially as "Attachment 1," "Attachment 2," etc.

Note: To conserve space in these guidelines, only Attachment 1 is included with this report.

False troughs are caused by minute intermittent leaks in the total gas purge system, resulting in a temporary loss of pressure. They appear on the chart as a comparatively rapid drop or rise in water level followed by a slow recovery. Intermittent leaks are more likely to occur at high water stages, with the result that crests are recorded one or two metres below their true stage, although this was not apparent in our tests. Very small intermittent leaks can be extremely difficult to locate.

The leak is most likely within the manometer. Tests of the digital recorder show it to be accurate, while the pre-test pressure check of the gas delivery system, and the soap test of the connections, demonstrate that there are no leaks between the manometer and the underwater orifice.

Conclusions

(15) Our tests show that the erratic water stage readings recorded at AWCS site 24 were caused by a tiny, undetectable internal gas leak within the water stage manometer.

Tests performed by   Approved by

Carole Winterton   *FL Cairns*

Carole Winterton   Frederick L. Cairns, P. E.
Lab Technician   Supervisor, Tests & Procedures

4

Attachment 1

⑯   Extract from Manufacturer's Manual for
Water Stage Manometer Model WSM

8. Instructions for Purging the System

To purge (introduce nitrogen gas into) the system, proceed exactly in the following sequence:

8.1   Check that the following valves are closed:

Valve	Rotation
Feed pressure adjustment screw	Fully CCW
Feed rate adjustment needle valve	Fully CW
Manometer shut-off valve	Fully CW

8.2   Turn the bubble tube shut-off valve fully CW (open).

8.3   Turn the nitrogen cylinder valve fully CCW (open).

8.4   Rotate the feed pressure adjustment screw slowly CW until the pressure gauge reads 280 kPa.

8.5   Rotate the feed rate adjustment valve slowly CCW until bubbles can be seen flowing in the manometer sight feed. Adjust the valve until the flow rate is 15 bubbles per minute.

8.6   Check that the flow rate at the orifice is 7 or 8 bubbles per minute.

Academic Laboratory Reports

Laboratory reports written in an academic setting use the same writing compartments as those written in industry (see Figure 5–1 on page 57), but there is a shift in purpose and emphasis. An industrial laboratory report responds to a specific request or demand and so answers a question or meets a stated need. An academic laboratory report is used by students to prove a hypothesis or test a theory, or as a vehicle for helping them learn how to perform a particular test or understand a process or procedure. Normally it does not respond to a tangible demand (other than a professor's request) or meet a specific need. It may, however, answer a question.

The science and engineering departments at each university or college often specify differing requirements for lab reports, which makes it difficult for us to specify exact writing compartments. Those described here offer the most generally accepted approach.

- A **Summary** (sometimes called an **Abstract**), which includes a brief statement of the purpose or objective of the tests, the major findings, and what was deduced from them.
- A more detailed statement of **Purpose** or **Objective**, plus other pertinent background data. (This writing compartment may be combined with the Summary if there is little background information.)
- An **Equipment Setup** compartment, which provides a list of the equipment and materials used for the tests and a description and illustration of how the equipment was interconnected. (If there is a series of tests requiring different equipment configurations, a full list of equipment and materials should appear here. A description and illustration of each setup should then be inserted at the beginning of each test description.)
- A **Method** compartment, containing a step-by-step detailed description of each test, similar to the Test Method section of an industrial laboratory report. Attachments may be used for lengthy procedures or process information.
- A **Results** compartment, giving a statement of the test results, or findings evolving from the tests.
- A detailed **Analysis** of the results or findings, their implications, and what can be learned or interpreted from them.
- A **Conclusions** compartment, comprising a brief statement describing how the tests, findings, and resulting analysis have met the objective stated in the Objective or Purpose compartment.
- A **Data (Attachments)** compartment, which is placed on a separate sheet (or sheets). It contains data derived during the tests, such as detailed calculations, measurements, weights, stresses, and sound levels. Lengthy procedures or process descriptions are sometimes included as attachments.

If several tests were performed, and there were results from each, it may be better to have separate Equipment Setup, Method, and Results compartments for each test. The organization plan would then be:

Summary

Objective or Purpose

Equipment and Materials

Test No. 1:
       Equipment Setup
       Method
       Results

Test No. 2:
       Equipment Setup
       Method
       Results

Test No. 3:
       Equipment Setup
       Method
       Results

Analysis

Conclusions

Data (Attachments)

In particularly comprehensive lab reports, there may be analyses at the end of each test.

CHAPTER 6

Investigation and Evaluation Reports

An investigation report describes a problem or situation that has been investigated, examines methods for correcting the problem or improving the situation, and usually suggests what should be done. If the report evaluates alternatives (such as an examination of several sites for locating a fast-food restaurant), it may be called an evaluation report; if it examines the practicability of doing something new or different, it may be called a feasibility study.

Investigation reports can be as short as only one or two pages, but more often they are much longer, ranging from 3 to 30 or 40 pages. Shorter reports are normally prepared as interoffice memorandums or letters, while longer reports usually have a title centered at the head of the first page (i.e. they become semi-formal reports and often are preceded by a cover letter). Some are even given the full formal report treatment shown in Chapter 8.

A very short investigation report is discussed in Chapter 4. It has five compartments, which are shown in Figure 4–3 on page 52. For longer investigation reports the compartments are expanded to introduce subcompartments as shown in Figure 6–1. We will identify and describe these compartments and subcompartments in greater depth in the comments for the four-page investigation report that starts on page 73.

Figure 6–1　Writing compartments for a long investigation report.

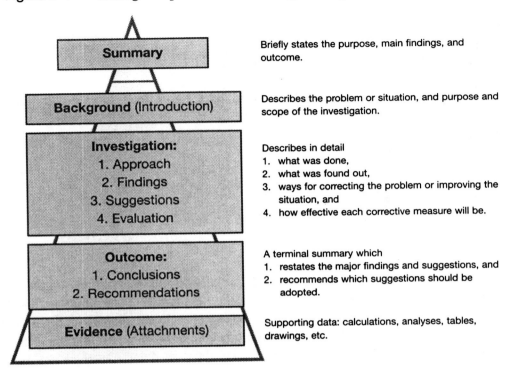

Summary

Briefly states the purpose, main findings, and outcome.

Background (Introduction)

Describes the problem or situation, and purpose and scope of the investigation.

Investigation:
1. Approach
2. Findings
3. Suggestions
4. Evaluation

Describes in detail
1. what was done,
2. what was found out,
3. ways for correcting the problem or improving the situation, and
4. how effective each corrective measure will be.

Outcome:
1. Conclusions
2. Recommendations

A terminal summary which
1. restates the major findings and suggestions, and
2. recommends which suggestions should be adopted.

Evidence (Attachments)

Supporting data: calculations, analyses, tables, drawings, etc.

Semiformal Investigation Report
Study of High Gas Consumption
(Pages 73 to 79)
Comments

Throughout his report author Tod Phillips keeps his readers firmly in mind. He is aware that, although he has done his investigation and is preparing his report for the Marsland Construction Company, his real readers are the homeowners, the Parsenons. For their sake he develops his case carefully and uses language and terminology they will readily understand.

If the Marsland Construction Company had simply asked Tod to investigate the cause of high gas consumption, and if he knew that only the company would use the information, then he could be much more direct. He could tell them in one sentence that he has checked the gas furnace, hot-air ducts, gas flow meter, and insulation and found them all to be satisfactory. Then he could go straight into his comparative analysis of gas consumption records and thermostat settings. The report then would be no longer than one or two pages.

But Jack Marsland has briefed Tod that the report is needed specifically for the Parsenons, who have complained of high gas consumption ever since the house was new and they first occupied it. Over the past three years, Marsland Construction Company has done numerous checks, made many adjustments, and found little wrong with the house. And yet the Parsenons have still complained. Finally, Jack Marsland called in a firm of professional consultants (H. L. Winman and Associates, who assigned the project to Tod Phillips) and asked them to carry out an independent study to identify the cause of the problem and recommend how it could be resolved. Jack Marsland could then present the results to the homeowners to persuade them that his construction company has done all it can to rectify the problem. As Jack Marsland told Tod Phillips: "Whatever you discover is wrong, I'll fix. If the construction is not at fault, then I can use your report to get the Parsenons off my back!"

Some comments on the implications of addressing the report directly to the home-owners rather than to the construction company follow the report (on page 82).

The comments that follow are cross-referenced to the corresponding numbers marked beside the investigation report.

(1) A title or main heading should be noticeable and informative—it should describe what the report is about. Too often, a report title can leave readers wondering what the report covers; for example, "Gas Consumption Investigation Report" would have been an inadequate title.

Although Tod Phillips could easily have prepared this report in letter form, he chose to write it as a semiformal report because he would then appear much more objective. He decided he could make any personal comments in a cover letter.

(2) A **Summary** carries the report's highlights, stated as briefly as possible. For an investigation report the Summary should cover

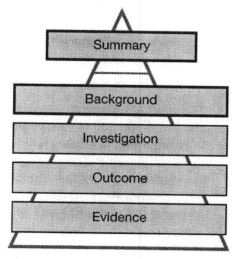

- the purpose of the investigation ("*Our investigation of heating fuel consumption ...*"),

- the major findings ("*... consumption ... slightly higher than in homes of comparable size and construction ...*"), and

- the major conclusions and/or recommendations ("*... can reduce consumption ... by maintaining the home at a slightly lower temperature ... and by insulating ...*").

In a long report like this, the summary is often written after the rest of the report. The report writer can then extract the highlights from the Background, Facts & Events, and Outcome compartments.

(3) When headings are used in a report, the **Background** compartment is titled "Introduction." An Introduction usually contains three pieces of information, not necessarily in this order:

1. The events leading up to the investigation.
2. The purpose of the investigation.
3. The scope of the investigation.

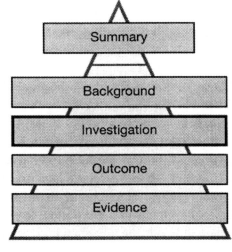

(4) The **Investigation** compartment starts here with the **Approach**, which explains that the study had two phases and so prepares the reader to find these two phases treated separately. The heading immediately following this paragraph tells the reader that phase one is about to start.

(5) Not all the headings exactly parallel the compartments used to write the report. To place a heading at the beginning of each compartment would have made the report too rigidly structured. Tod Phillips used the compartments to ensure that he was organizing his report properly, and then later inserted headings where they would help readers see the logical divisions of information. Headings that most often parallel the report writing compartments include:

Summary (Main Message)

Introduction (Background)

Conclusions and Recommendations (Outcome)

Attachments (Evidence)

(6) This short paragraph continues the **Approach** subcompartment by identifying what physical checks will be carried out in the Parsenon home. An "overview" paragraph following a section heading is useful because it prepares the reader to expect a certain narrative sequence. Later in the report—at (9) —a second Approach paragraph describes how the comparative analysis was carried out. Although many investigation reports have a straight-through discussion, in which the *whole* approach is presented first and is followed immediately by *all* the findings, Tod has chosen to use a two-stage method because he has two distinct aspects to deal with: the physical check of the home, and the comparative analysis of gas consumption records.

(7) The **Findings** subcompartment starts here and continues through four paragraphs. [A second set of findings appears later in the report, at (11).]

H. L. Winman and Associates
Professional Consulting Engineers

(1) Study of High Gas Consumption at 1404 Gregory Avenue

Summary

(2) Our investigation of heating fuel consumption in Mr. and Ms. R. M. Parsenon's home at 1404 Gregory Avenue shows consumption to be 6.4% higher than in homes of comparable size and construction. We suggest that Mr. and Ms. Parsenon can reduce consumption to a normal or even lower level by maintaining their home at a slightly lower temperature and by insulating the upper 4 feet of the basement walls. They could achieve even greater fuel saving by installing additional insulation in the ceiling and walls.

Introduction

(3) The investigation followed lengthy correspondence between Marsland Construction Company, who built the one-storey home in 1996, and the homeowners, who have continually reported excessively high fuel consumption and uncomfortably cold floors. In a letter dated February 16, 2001, Marsland Construction Company authorized us to carry out an independent study to determine the extent of high fuel consumption, identify the cause, and suggest remedial measures.

(4) We divided the study into two phases: a physical check of the building and its heating system, and a comparison of gas consumption in the Parsenon home with gas consumption in similar homes.

Checks of Building and Heating System

(5)
(6) Our examination of the Parsenon home involved checks of the gas furnace and hot air ducts, the gas company's flow meter, and the exterior walls and insulation.

(7) Montrose Heating and Supply examined the gas furnace and hot air ducts and found them to be in good condition. The one exception was severe corrosion of the humidifier plates in the gas furnace. Although this condition could cause lower humidity in the home, it would not affect the temperature. Montrose Heating installed new humidifier plates and left a spare set with the homeowners.

Montrose Gas Company checked the gas flow meter but also found no fault. They reported that the meter was replaced twice in the past 12 months at the homeowners' request, and on both occasions no fault was found.

Our examination of the exterior walls revealed there are no poorly insulated areas and no ill-fitting doors or windows. Close examination of the insulation showed that

- the ceiling has 8 inches of wood chips covered by a further 6 inches of Qualathane foam, which together provide an R40 insulation factor,

⑧ Tod has found that the problem lies entirely with the homeowners and not with the construction company. So he has to tell the Parsenons something they don't want to hear—that the builder is blameless and that they have to take whatever corrective action is necessary. Consequently he has to *persuade* them to accept his findings. His careful accumulation of evidence in this first set of investigation findings does exactly that. He cannot dismiss the insulation and heating equipment in just a few words, because the Parsenons most likely suspect that these are the cause of their problem. So he carefully examines each aspect, quotes *facts* to prove there are no faults, and does not allow his personal opinions to intrude. By the end of this section there must be no doubt in the Parsenons' minds that the fault lies elsewhere.

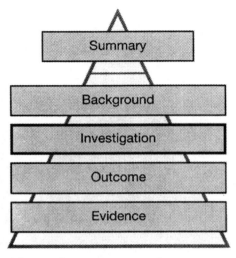

⑨ This is the second part of the **Approach** subcompartment.

⑩ A reader should be able to read a whole report right through without having to turn to the attachments, but should always be informed at appropriate places that supportive evidence is attached and where to find it.

⑪ The second set of investigation findings starts here. The results of this part of the investigation are based on evidence supplied in an attached comparative analysis. Tod presents his findings here without discussion; in effect he is saying to readers: "This is what I found out, and you can verify it by referring to the comparative analysis." His use of brief subparagraphs (point form) helps maintain a detached, almost clinical, presentation of these facts.

⑫ In the **Suggestions** subcompartment a report writer can offer a single suggestion, several suggestions, or alternative suggestions. Tod presents two main suggestions, and implies that one should be adopted while the other is optional.

⑬ The **Evaluation** subcompartment allows a report writer to shed some of his or her impartiality. Readers like to form their own opinions of the validity of each suggestion, but need a convincing, logical, rational evaluation on which to base them. They depend on the report writer to evaluate factors such as

- feasibility,
- suitability,
- simplicity,
- effectiveness, and
- cost.

(8)
- the walls are filled with Fiberglass batts and have an R24 insulation factor, and
- the basement is unfinished and uninsulated.

This level of insulation meets current state standards: i.e. a minimum of R40 in the ceiling and R20 in the walls.

(9) **Comparison with Other Homes**

As our checks of the Parsenon home showed no significant cause for the reported high gas consumption, we decided to compare consumption in the Parsenon home with consumption in two groups of single-level homes of comparable age and size. They comprised

- four identical homes built by the same contractor; these were Gregory Avenue numbers 1396, 1399, 1407, and 1410,
- four homes built at the same time, but by other contractors; these were Gregory Avenue numbers 1506, 1515, 1524, and 1581.

(10) In each case we obtained permission from the homeowners to examine and quote from their consumption records for the calendar year 2000. We also asked homeowners to inform us if they had installed additional insulation since their home had been built, and the setting at which they kept their thermostats. The results are shown in the attached table.

Examination of the comparison table shows that

(11)
- gas consumption in the Parsenon home for 2000 was 55.1 MCF (55,100 cubic feet) or 6.4%, greater than the average of the eight other homes we evaluated,
- the thermostat in the Parsenon home was set 3°F higher by day, and 8°F higher at night, than the average setting for the eight other homes,
- gas consumption for the six homes in which the thermostat setting was lowered at night was consistently lower than the consumption in the three homes in which the thermostat setting was not lowered.
- gas consumption in Marsland-built homes was comparable to that in homes built by other contractors, and
- two homes in which additional insulation has been installed consumed significantly less gas than homes which have only their original insulation.

We also checked whether the Parsenon residence had other gas-fired equipment such as a clothes drier or fireplace, but it has neither. However, it does have a gas-fired hot water heater, but this is common to all nine homes we evaluated.

Methods for Reducing Gas Consumption

(12) We believe Mr. and Ms. Parsenon can reduce gas consumption in their home to an average level, or even slightly better than average, at little or no cost. This reduction can be achieved by lowering the thermostat setting from its present constant 73° to 70° during the day and 64° at night. They could achieve a further significant reduction, but at extra cost, by installing additional insulation in the ceiling, walls, and basement.

(Sometimes a report writer may also have to evaluate the effect of taking *no* action, i.e. adopting none of the suggestions.)

Tod Phillips has to convince his secondary readers (the Parsenons) that they must take at least one step, and preferably two, to resolve their problem. He tries to persuade them to accept his suggestions by

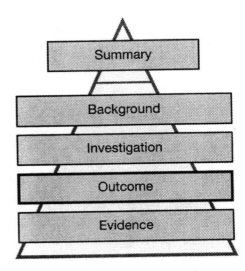

- presenting evidence to show they have to lower the temperature in their home (he quotes federal specifications to demonstrate what gas saving they should be able to achieve and then reinforces his case by referring to local residents who are achieving such savings),

- suggesting they would be wise to insulate at least part of their basement (he acknowledges their problem of cold floors and then points to what someone else has done), and

- explaining how they could gain additional savings by installing more insulation in the walls and ceiling of their home (he recognizes this would be costly and so presents it only as an option).

It would be difficult to refute the validity of Tod's evaluation. It is carefully developed, so that it carries readers logically from one point to the next. By now, Tod's readers should feel they *know* what the conclusions and recommendations are going to be.

(14) Tod has written a straightforward evaluation because he is offering several suggestions, which will have a cumulative effect. His readers have to decide how many of his suggestions should be adopted, rather than choose between them. He might have arranged his information differently if there had been alternative suggestions. For example, suppose that you have investigated several word processing systems to identify which would be most suitable to install in your office. Now you have to write an investigation report to management, in which you will recommend the best system. There are two methods you can use for arranging the **Suggestions** and **Evaluation** subcompartments:

- You can present all the suggestions first, and then evaluate them (as Tod has done).

(13) Although it is difficult to predict an exact saving in gas consumption, Federal Insulation Specification FIS-2820/78 suggests that for every 1000 square feet of floor space, a 1° reduction in house temperature conserves about 15.4 MCF of gas annually (MCF = 1000 cubic feet). In the Parsenon residence, a 3° reduction in temperature should therefore cut gas consumption by 46.2 MCF annually. We also suggest that a further 8 to 10 MCF of gas could be saved annually by reducing the temperature to 64° at night. These estimates seem to be borne out by the consumption figures for the homes at 1407, 1506, and 1581 Gregory Avenue, in which lower temperatures are maintained without additional insulation.

Ms. Parsenon has commented that they are forced to keep their home temperature at 73° because at any lower temperature the floor is too cold. We asked the owners of an identical home at 1410 Gregory Avenue, who maintain a 69° temperature in their home, if their floors are similarly cold. Their experience offers a probable solution. After their basement was insulated they found the floors to be considerably warmer and that they could reduce the thermostat setting from its previous 72° to 69° without discomfort.

(14) The basement at 1410 Gregory Avenue has been fully insulated. We believe it would be necessary to insulate only the upper 4 feet of a basement to obtain a similar result, because below the 4 foot level the walls are insulated naturally by the surrounding soil. Basement insulation is easily applied by nailing or gluing styrofoam panels directly onto the walls. We estimate that to insulate the upper 4 feet of the basement would cost $1250 if the work is done by a contractor, or $525 if it is done by the homeowners.

(13) Installing additional insulation in the ceiling and walls would achieve even greater savings in gas consumption, and also enhance the comfort level at the suggested thermostat settings. In 2000, the fully insulated home at 1515 Gregory Avenue consumed 55.1 MCF less gas than the average for the eight homes we evaluated, and 110.3 MCF less gas than the Parsenon home. Significantly, it is also the largest home in the group.

Because the cost of insulating the ceiling and walls, and even the basement, of their home will depend on the quality (R factor) of insulation they choose, we suggest that Mr. and Ms. Parsenon obtain advice and an estimate from a recognized insulation contractor.

Conclusions

(15) Our study shows that gas consumption in Mr. and Ms. Parsenon's home is 6.4% higher than average, and that this high consumption is attributable primarily to the home being maintained at a higher-than-average room temperature.

Consumption could be reduced to average for this type and size of dwelling by lowering the thermostat setting by 3°F during the day and by 8°F at night, and by maintaining an optimum humidity level. It could be reduced to better than average by also installing additional insulation.

Suggestions:
   System A
   System B
   System C

Evaluation:
   System A
   System B
   System C

- You can present each suggestion in turn and then immediately evaluate it.

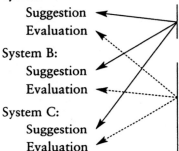

System A:
   Suggestion
   Evaluation

System B:
   Suggestion
   Evaluation

System C:
   Suggestion
   Evaluation

A description of the system
and its main features

An analysis of its
main advantages and
disadvantages, and its
suitability for your office

If you are evaluating products or alternative corrective measures, only one of which will be selected, try to maintain your objectivity throughout the evaluation. That is, compare each product or corrective measure against a set of evaluation criteria rather than compare them against one another. If you plan to present all the suggestions first, then establish and prove your evaluation criteria *after* you have presented your suggestions but before you evaluate them. If, however, you plan to present each suggestion independently and will immediately evaluate it, then establish and prove your criteria *before* you present the first suggestion. (We describe in detail how to write a comparative analysis on page 140.)

(15) The conclusions and recommendations form the report's **Outcome** compartment. They present the results of the investigation, suggest what needs to be done, and sometimes recommend what action the reader should take.

The conclusions should repeat the main features of the **Findings** subcompartment. They must never introduce any information or ideas that have not been discussed previously in the report.

(16) Recommendations must be strong and definite. They should be stated in the active voice rather than in the bland passive voice. That is, they should say "I recommend ..." or "We recommend ..." and *not* "It is recommended that ... ." And they should never recommend action that has not already been discussed in the report.

Recommendations

(16) We recommend that Mr. and Ms. Parsenon reduce the temperature in their home to 70°F during the day and 64°F at night, increase the floor temperature by insulating the upper 4 feet of the basement walls, and maintain constant humidity by replacing the humidifier plates in the gas furnace at least once a year.

If this does not achieve the desired reduction in gas consumption, then we suggest that Mr. and Ms. Parsenon install additional insulation in the walls and ceiling of their home.

(17)

T. E. Phillips

Tod E. Phillips, P.E.
March 16, 2001

4

(17) The signature at the end of the report is optional. The date the report is issued should be included, however, either here or immediately after the report title.

(18) Attachments (sometimes called Appendixes) hold factual evidence that supports statements made in the report but that is too comprehensive, complex, or detailed to be included with the main narrative. They may be calculations, analyses, cost estimates, drawings, photographs, plans, or copies of other documents. They form the **Evidence** compartment of the report writing pyramid. Certain "rules" apply to attachments:

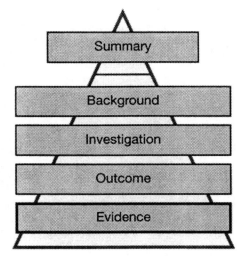

- Readers should not have to turn back to them as they read the report (although they may choose to refer to them later). This means that the most significant features of an attachment must be quoted in the investigation narrative.

- Every attachment must be referred to in the report narrative. It must serve a specific purpose and never be included simply because it contains possibly useful information.

- The attachments should appear in the order in which they are referred to in the narrative. If there are several attachments, they should be numbered Attachment 1, Attachment 2, etc. (Or, if they are referred to as appendixes, as Appendix A, B, C, etc. Appendixes are more often used with formal reports than with semiformal reports.)

Comparison between Semiformal and Letter-Form Investigation Reports

A semiformal investigation report is generally longer, seems to have more dignity, and appears to be more formal than a letter. Its title is centered at the top, its contents are clearly divided into compartments, each preceded by a heading, and its language is usually a little less personal. It can be a useful way to convey information when the contents of a report are meant to influence a third party. (In the example, the report is written for the Marsland Construction Company, but its results are really intended for the homeowners.)

Because it is not addressed to a particular reader, a semiformal investigation report usually needs to be accompanied by a cover letter.

(18)

Attachment

Comparison of Gas Consumption
in Nine Gregory Avenue Homes
for Calendar Year 2000

House No.	Year Built	Size (sq ft)	Consumption (MCF)	Thermostat Day (°F)	Thermostat Night (°F)	Additional Insulation
Parsenon Residence:						
1404	1996	1004	923.6	73.0	73.0	None
Group 1:						
1396	1996	1004	870.6	72.0	64.0	None
1399	1996	1004	894.4	70.0	70.0	None
1407	1995	1004	880.9	70.0	63.0	None
1410	1996	1004	841.3	69.0	69.0	Basement
Average:		**1004**	**871.8**	**70.2**	**66.5**	
Group 2:						
1506	1994	966	868.9	71.0	60.0	None
1515	1993	1080	813.3	70.0	60.0	Full
1524	1992	980	900.3	72.0	72.0	None
1581	1994	966	877.5	70.0	64.0	None
Average:		**998**	**865.0**	**70.7**	**64.0**	
Average of 8 homes:		**1001**	**868.4**	**70.4**	**65.2**	

LEGEND: Group 1:   Four identical homes built by Marsland Construction Company.
Group 2:   Four nonidentical homes built by other contractors.
MCF: 1000 cubic feet.

The cover letter serves two purposes:

- It very briefly summarizes the results presented in the report.
- It provides a place for report writers to make comments they might prefer not to insert in the report.

The cover letter that accompanied Tod Phillips's report for the Marsland Construction Company does both:

> Dear Mr. Marsland:
>
> Our investigation at 1404 Gregory Avenue shows that Mr. and Ms. Parsenon's complaint of high gas consumption is partly justified. We find, however, that it is caused by the high temperature at which the Parsenons keep their home rather than by faulty heating equipment or inadequate insulation.
>
> We hope the attached report will help you convince Mr. and Ms. Parsenon to take the appropriate steps necessary to bring gas consumption down to an acceptable level.
>
> Sincerely,
>
> Tod Phillips

A letter-form investigation report, however, does not normally exceed two or three pages, addresses the recipient personally, and may still contain headings. If Tod Phillips's report had been written *as a letter* to the Marsland Construction Company, the first paragraph (the Summary) probably would not have changed. However, subsequent paragraphs would have contained more personal pronouns, such as "you," "your," "we," and "our." For example, the second paragraph—at ③ —probably would have looked like this:

> Our investigation followed your lengthy correspondence with Mr. and Ms. Parsenon regarding their complaints of high gas consumption and cold floors. In your letter of February 16, 2001, you authorized us to determine the extent of high fuel consumption, identify the cause, and suggest possible remedies.

This "personalization" of a report's language becomes even more noticeable when the person to whom a letter report is addressed is an individual citizen rather than a company. If Tod Phillips had addressed his report directly to Mr. and Ms. Parsenon, rather than to the construction company, even its summary would have been more personal:

Dear Mr. and Ms. Parsenon:

We have investigated the reported high heating fuel consumption in your home and have found it to be 6.4% higher than in homes of comparable age, size, and construction. Our checks revealed neither faults in the heating system nor inadequate insulation of your home during its construction. Comparison with similar homes, however, showed that the high gas consumption is probably caused by the higher-than-average temperature at which you maintain your home.

We believe you could reduce gas consumption to a normal or even lower level by maintaining a slightly lower temperature in your home during the day, lowering the thermostat to about 64°F at night, and insulating the upper 4 feet of the basement walls. If you want to obtain even greater fuel savings, then we suggest you consider installing additional insulation in the ceiling and walls.

Our investigation was requested by ... (etc.).

CHAPTER 7

Suggestions and Proposals

Proposals vary from short memorandums to multivolume hardbound documents. Those discussed here are the shorter, less formal versions, ranging from a single page to about 10 pages.

There are three types of proposals:

- Informal suggestions
- Semiformal proposals
- Formal proposals

A **Suggestion** offers an idea and briefly discusses its advantages and disadvantages. (For example, a supervisor may suggest to a department manager that break times be staggered, to avoid line-ups at the mobile refreshment wagon.) Most suggestions are internal documents and are written as memorandums.

A **Semiformal Proposal** presents ideas for resolving a problem or improving a situation, evaluates them against certain criteria, and often recommends what action should be taken. (For example, a supervisor may propose to management that steps be taken to overcome production bottlenecks in the company's packing department. He or she might suggest introducing new packaging equipment, discuss various alternatives such as hiring additional staff or embarking on a training program, and then recommend the most suitable approach.) A semiformal proposal may be written as a memorandum, a letter, or in semiformal report format.

A **Formal Proposal** describes an organization's plans for carrying out a large project for a major client, such as the government. It is a substantial, often impressive document, which describes in considerable detail what will be done, how and when it will be done, why the organization has the capability to do the work, and what it will cost. Such proposals are often prepared in response to a Request for Proposal (RFP) and are almost always submitted as bound books similar to a formal report. In extreme cases they may run to several volumes.

We are providing writing plans for all three types of proposals and examples (with accompanying comments) for the informal suggestion and two semiformal proposals.

Informal Suggestions

The writing plan for an informal suggestion is illustrated in Figure 7–1.

- The **Summary** states very briefly what the proposer wants to do or wants done.
- The **Background** compartment describes the present situation.
- The **Details** compartment has two components:
    1. A **Suggestion** section, which outlines the suggested changes or improvements, and describes why they are needed.
    2. An **Evaluation** section, which identifies what effect the suggested changes or improvements will have and categorizes them into advantages and disadvantages (sometimes called "Gains" and "Losses").

Figure 7–1   Writing plan for an informal suggestion.

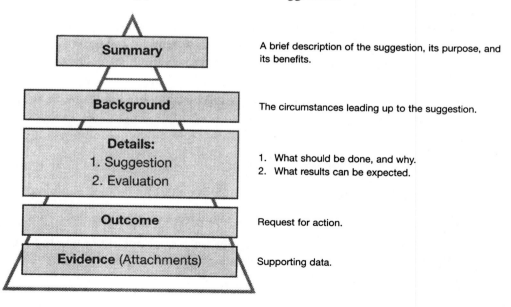

- The **Outcome** compartment identifies what action needs to be taken. It can either request approval for the author to implement his or her suggestion or identify who should take the necessary action, and possibly describe when and how it should occur.

- An optional **Evidence** compartment contains supporting data such as cost estimates, records, plans, and sketches.

Informal Suggestion
Proposal for a Study
Comments

When an informal suggestion is as short as Angie Morton's (opposite), the writing compartments may not be as clearly evident as those used to shape a longer memo, letter, or report. Nevertheless they are still there, contributing to the overall organization of her suggestion.

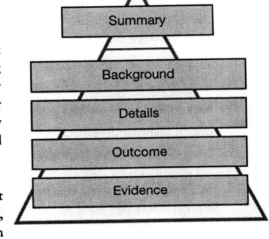

① Angie uses a cause-and-effect approach for her **Summary Statement**, first very briefly identifying her reason and then stating what she would like to do in general terms.

② This paragraph is the **Background** compartment; it amplifies the reason stated in the Summary.

③ The **Details** paragraph contains a specific **Suggestion** (in sentence 1) and a short Evaluation, which spells out what is to be gained (sentence 2) and the cost (sentence 3). Angie keeps the evaluation short by referring to details in an attachment (i.e. in the **Evidence**). To conserve space the attachment is not included here.

④ In the **Outcome** compartment Angie states specifically what she wants to do and requests approval (action).

Semiformal Proposals

A semiformal proposal is more comprehensive than an informal suggestion:

- It usually deals with more complex situations, such as a problem or unsatisfactory condition.

- It discusses the circumstances in more detail.

Electro-Mechanical Engineering Services Inc.
Interoffice Memorandum

To: Frances Kelvin, Manager
Field Engineering Services

Date: February 15, 2001

From: Angie Morton, Coordinator
Technical Services   *AM*

Subject: Proposal for a
Computer Study

① We have included $24,000 in this year's budget for the purchase of eight portable computers for our field engineers. To ensure we buy not only the best but also the most suitable computers, I propose we engage a consultant to identify our exact needs.

② The portable computer field is advancing so rapidly that it's difficult to assess which model and what features we require. Some of our field engineers and several sales representatives have recommended various existing makes and models, while others have recommended we should wait for more advanced models currently being developed.

③ I have spoken to Gavin Gray at Antioch Business Consultants who, as his attached letter describes, can assess our existing and potential needs in the light of current and upcoming portable computer technology. He will evaluate the alternatives available and recommend an optimum system or model, and will complete his evaluation within two weeks of our authorization to proceed. He has quoted a firm price of $1025 plus tax to undertake the study, which can be drawn from the portable computer acquisition budget.

④ I suggest we authorize Antioch Business Consultants to carry out the study and request your approval to go ahead.

- It establishes criteria (guidelines) for any proposed changes.
- It frequently offers alternatives, rather than a single suggestion.
- It analyses the proposed alternatives in depth.
- It has a more formal appearance.

Two approaches are described here: one for proposals that suggest how to correct a problem or improve a situation; the other for proposals that offer a service to a client.

Proposals That Present an Idea

The writing compartments for a semiformal proposal that examines a problem and suggests a solution, or offers an idea for consideration, are shown in Figure 7–2.

- The **Summary** briefly describes the main highlights of the proposal, drawn mostly from the Background, Solutions, and Outcome compartments. If headings are used in the proposal, this compartment is preceded by the word SUMMARY or ABSTRACT.

Figure 7–2   Writing plan for a semiformal proposal that presents an idea.

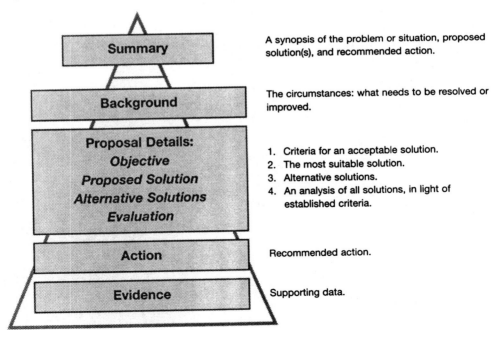

- The **Background** compartment introduces the problem, situation, or unsatisfactory condition and outlines the circumstances leading up to it. It may be preceded by the heading INTRODUCTION.

- The **Objective** subcompartment defines what needs to be achieved to resolve the problem, and establishes criteria for an optimum, or best, solution. This information may be included as part of the INTRODUCTION or preceded by a heading of its own (such as REQUIREMENTS or CRITERIA).

- The two **Solutions** subcompartments describe various ways the problem can be resolved or the situation improved. Each alternative should include

    - a description of the solution,
    - the result or improvement it would achieve,
    - how it will be implemented,
    - its advantages and disadvantages, and
    - its cost.

  Ideally, the alternative solutions will be arranged in descending order of suitability. These two subcompartments may be preceded by a single heading, such as **Methods for Increasing Productivity**, or by several descriptive headings.

- The **Evaluation** subcompartment analyzes and compares the alternative solutions, with particular reference to the criteria established in the Objective subcompartment. It may briefly discuss the effects of

    - adopting the proposed solution,
    - adopting each of the alternative solutions, and
    - adopting none of the solutions (i.e. taking no action).

- The **Outcome** compartment recommends what action should be taken. It should be worded using strong, positive terms and, if headings are used, the section should be preceded by the single word RECOMMENDATION (or RECOMMENDATIONS).

- The **Evidence** compartment, if used, contains detailed analyses, test results, drawings, etc, which support and amplify statements made in the previous compartments. It is usually preceded by the heading ATTACHMENTS or APPENDICES.

The semiformal proposal on the following pages shows how these compartments can help organize information into a coherent, convincing, persuasive document.

Semiformal Proposal No. 1
Proposal to Install
Videoconference Facilities
in the Three Capilano
Group Divisions
Comments

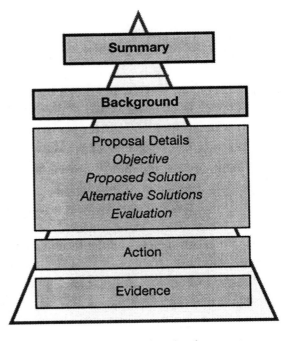

① Leo Cheng has used effective Information Design to present his proposal to the Board of Directors of The Courtland Group—an 8700-employee insurance company operating near both coasts and in Minnesota. Leo is the company's Manager of Training and Communication. The two-column approach he has used creates attractive, open "white space" that appeals to readers. The open column on the left side also encourages them to make notes in the margin. (For more about Information Design, refer to Chapter 9.)

② Although he has prepared the proposal for Dana Livingstone—the Director of Corporate Affairs—he knows she will take it to the next Board meeting and present it to the Board members for approval. Therefore, he is aware of them really as the primary audience, because they will decide whether to go ahead with the project.

③ Leo's **Summary** is brief yet complete. It identifies why the proposal has been written, what is being proposed, what it will cost, and how quickly the cost will be recovered.

④ The **Background** compartment starts here. Some Board members will know more about the company's history than others. Since the company is located in California and has only recently taken over a company in Connecticut, Leo describes corporate history in sufficient detail in a page and a half so that all members will understand the rationale for the proposal he is about to make.

① Proposal to Install Videoconference Facilities
in the Three Courtland Group Divisions

prepared for

② Dana L. Livingstone
Director of Corporate Affairs
The Courtland Group

③ *To increase communication between Divisions, and concurrently reduce the cost of Board meetings and training, we propose that The Courtland Group invest $223,200 to install three state-of-the-art videoconference centers, one each in the company's offices in San Francisco, CA, St. Cloud, MN, and New Haven, CT. The capital cost will be recovered in 13.3 months.*

④ **History of Courtland Life and Trust, and The Courtland Group** Since its inception in San Francisco in 1894, Courtland Life and Trust Company's management philosophy has resulted in the company enjoying significant growth. This philosophy has embraced two key factors:

1. Effective communication between the Board of Directors and company management.

2. Comprehensive and continuing training for the company's sales force and operating staff.

This philosophy worked well while Courtland Life and Trust operated almost entirely in California. However, when in 1998 the company acquired Avenue West Insurance Group in Minnesota and Wisconsin, and then in 2000 purchased Schönberg Mutual in Connecticut, communication between Board members became more difficult and more expensive, and new strategies had to be implemented to maintain continuity of training throughout the organization.

To recognize the broader nature of the company's business, on January 1, 2001 the company changed its corporate name to The Courtland Group. The company then had 8700 employees: 3700 in the Pacific Division (which included headquarters' staff) based in San Francisco; 2400 in the Central Division based in St. Cloud; and 2600 in the Atlantic Division based in New Haven.

Amalgamating just two companies, each with its own corporate culture and methods of operation, inevitably poses problems. Amalgamating *three* organizations, all well established and operating in widely diverse parts of the continent, is even more difficult.

1

⑤ The longer a proposal is, the more need there is to insert headings to help steer readers through the information. This is true for a proposal written in either the traditional format or the Information Design format.

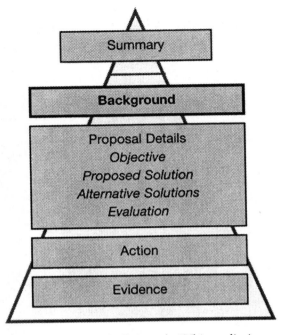

You can check the suitability of your headings by listing them on their own to see if there is good continuity. They should show the route you have taken and the natural flow of information that evolves from it. If they don't, you need to re-examine your headings to see if they are appropriate. Leo's headings, extracted from his proposal, are summarized on page 96.

⑥ The **Proposal Details** start here. Leo purposely starts with the words "This preliminary proposal..." because he wants to send a signal to his readers that he has yet to undertake a detailed study and write a definitive proposal that will provide precise descriptions of the renovations, furniture, equipment, and costs. Nevertheless, he has researched the situation and the alternatives in sufficient depth to be able to provide a reasonably accurate (although conservative) cost estimate.

⑦ By establishing clear-cut **Objectives** early in his proposal, Leo offers benchmarks against which the systems he will describe can be measured.

Two problems are particularly significant: maintaining effective communication among Board members, and introducing common operating standards throughout the company.

(5)

Communication Among Board Members

The 14 Board Directors meet four times a year. The annual general meeting is traditionally held in April at the company's headquarters in San Francisco. The three other meetings are each held in one of the Divisions: in New Haven, CT, in Santa Barbara, CA, and in St. Cloud, MN, in that sequence.

Because Board members come from all three Divisions, approximately 70% of the Board have to fly in to the meeting site. Travel, hotel, and meals cost about $37,500 per meeting, or $150,000 a year. We believe this cost can be reduced by 35% to 40% if videoconferencing technology is used for two of the meetings.

Common Operating Standards

The company has been gradually implementing common operating standards. This has proved to be a slow and expensive process, because teams of trainers have to travel to centers within each Division to meet with staff who will be implementing the new methods. For year 2001 this training cost will be $465,000 (for a nine-month period). For 2002 and 2003 we have budgeted $650,000 per year specifically to conduct training as corporate operating standards are brought on line.

We believe that by using videoconferencing we can reduce these costs by 20% to 25%.

Proposal Purpose

(6)

This preliminary proposal describes the advantages of designing and installing a state-of-the-art videoconferencing center at The Courtland Group's headquarters in San Francisco, and subsidiary videoconference centers in St. Cloud, MN and New Haven, CT. If management approves the concept, I will conduct further research and prepare an in-depth report describing the proposed system and its cost.

Objectives

(7)

If we are to adopt videoconferencing as a primary communication medium, we must be able to use the system equally well for Board of Directors' meetings and Training Department courses. At each location the system must offer

- a boardroom atmosphere, with seating for up to eight Board members,
- a classroom environment, with seating for up to 24 participants,
- connection to the Internet or Intranet, or both,
- high resolution/high definition video equipment, and
- the ability for seated Board members or training participants to become instant speakers without moving from their position.

2

⑧ The Proposal Details continue with the **Proposed Solution**, which continues for the next two and a half pages.

⑨ Each time Leo knows he is going to describe more than one aspect of a topic, he tells his readers exactly how he plans to present the information. For example, at the end of this introductory paragraph, he writes: "we plan to create a classroom first, and then fit the boardroom within it." He follows immediately by describing how the classroom will be designed, and then how the boardroom will be configured. By leading the reader to expect information to be

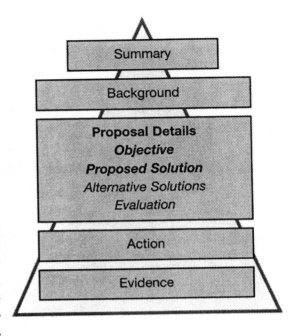

presented in a certain sequence, Leo shows he is managing his writing well.

Think of this as a miniature pyramid with a summary statement at the top and supporting details below it. Leo does the same in paragraph 1 on page 2 (it starts: "Two problems are particularly significant."), again on page 4 where he describes the technical facilities, and also on page 6 where he describes his evaluation of the alternatives.

⑩ Ideally, the text and the illustration that supports it should be on the same page, so readers do not have to flip back and forth. However, there may be occasions when this objective cannot be achieved.

⑪ There are two reasons for inserting illustrations into a proposal: they help reader understanding and they provide a more appealing overall image than just page after page of text. Compare, for example, page 4 of Leo's proposal with page 5. Although page 5 is pleasantly spaced and the side headings carry the reader easily through the narrative, it still has less visual appeal than page 4.

Since most teaching will be done from San Francisco, the head office videoconference center must also have a clearly defined teaching position and instructional facilities.

(8) **Videoconference Center Physical Design**

(9)

Each videoconference center has to be designed so that it can be converted readily from a meeting setting to a classroom setting. Its decor must also satisfy both functions. To do this we plan to create a classroom first, and then fit the boardroom within it.

Classroom

The classroom will contain three identical clusters of four tables formed into two rows, with each table having two seating positions. The clusters will be angled slightly toward the centre of the room, as shown in Figure 1, and will be separated from each other by a walk-through that is 30 inches wide.

(10)

(11)

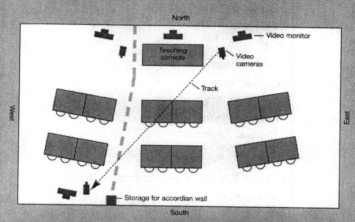

Figure 1. Design for classroom setting.

Boardroom

The boardroom will be situated at the west end of the classroom and will use the eight-seat cluster of tables at that end of the room. The four chairs between the tables will be rolled aside, and the two front tables will be on swivels so they can be rotated through 180° and slid toward and abutted against the two rear tables to form one large table. A light mahogany tabletop will be brought in to cover all four tables and so provide a dignified appearance. A folding wall will be pulled forward from the back of the room to separate the two remaining classroom clusters from the boardroom cluster, as shown in Figure 2.

3

Headings used in Leo Chang's proposal

Heading	Writing Compartment
History	Background
Communication Between Board Members	
Common Operating Standards	
Proposal Purpose	Objective
Objectives	
Videoconference Center Physical Design	Proposed Solution
Classroom	
Boardroom	
Decor	
Technical Facilities	
Video Cameras	
Video Monitors	
Control Console	
Individual Equipment	
Cost	
Feasibility of Renting Videoconference Rooms	Alternative Solutions
Convenience	
Cost	
Evaluation of Alternatives	Evaluation
Convenience	
Cost Comparison	
Current Method	
Renting	
In-House	
Recommendation	— Outcome
Attachment	— Evidence

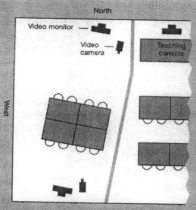

Figure 2. Design for boardroom setting.

Decor

The walls throughout the room will be covered with a patterned, sound-absorbent cloth that will be durable and yet suit a board-room setting. The folding partition will be covered with identical cloth. The floor will be covered with fitted carpet pieces that are 16 inches square to form an attractive pattern.

Technical Facilities

The technical facilities will comprise two components: (1) the video cameras and monitors, and (2) the microphones and technical equipment at the individual seating positions.

Video Cameras

There will be two video cameras. During teaching sessions, the two cameras will be high on the wall, one facing the eight seating positions at the east end of the room, and one facing the seating positions at the west end of the room (see Figure 1). Each will also cover the people seated in the middle cluster.

For Board meetings, the video camera at the east end of the room will travel on rollers along a ceiling-mounted rack and be repositioned on the back (south) wall at the west end of the room, where it will face the four Board members who will face south (see Figure 2).

Video Monitors

There will be three 48 inch video monitors mounted on the north wall, one facing each cluster of classroom participants. There will also be a similar-sized video monitor mounted on the south wall, beside the video camera position, for use during Board meetings.

Control Console

In the center of the north wall at the San Francisco videoconference center, there will be a teaching desk with a bank of four 14 inch video monitors for the instructor to cue input from the two cameras, a vcr, and an Internet connection point.

4

(12) Because this is only a preliminary proposal, with approximate dimensions and cost figures, it's sufficient for Leo to insert only basic illustrations that show the key points he describes. When he prepares his more detailed proposal, he will still provide basic illustrations embedded in the narrative, and support them with more complex, properly dimensioned illustrations in attachments.

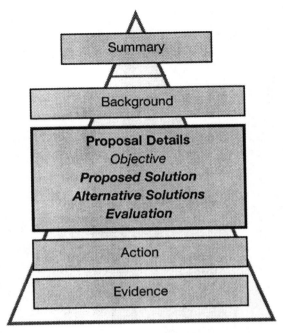

(13) Headings tend to be shorter when a writer uses a two-column style of Information Design, which can make it difficult to differentiate between primary headings and subordinate headings. There is insufficient space to use a significantly larger font, so Leo has chosen to use the same size font but to set subordinate headings in italics and indent them. All headings in his proposal are 12 pt—the same size as the main narrative—which simplifies typing.

(14) We asked Leo why he chose to write in the first person plural ("We") rather than the first person singular ("I").

> "Although I did the study and wrote the report," he said, "I'm really representing the Training and Communication Department. It sounds better for a proposal that will go to the Board of Directors. If I had been writing *only* to Dana Livingstone, I would have used 'I' and 'me'. "

(15) A synopsis of the total costs is sufficient here; readers will turn to the attachment later. This is an "open" table, which means its compartments are not separated by horizontal and vertical lines. (There is a "closed" table opposite comment 17.)

(16) The **Alternative Solutions** compartment starts here.

Individual Equipment

The individual positions will each have a surface-mounted microphone that protrudes only 4 inches above the desk surface, with a touch-sensitive on-off switch embedded in it. Touching the switch will enter the participant's position in an electronic cue. When the participant's turn occurs, the microphone will become "live" and the appropriate video camera will pan toward that speaker's position.

(14) At each position there also will be outlets beneath the table surface for plugging in a portable computer and a modem. We are anticipating that most future training will be done online.

(15) **Cost**

The total cost for building and equipping the three video-conference centers will be $223,200, broken down into four compartments:

Renovations to the three rooms	$77,000
Furniture (desks and chairs)	28,800
Purchase of technical equipment	90,400
Installation of equipment	27,000

(16) See the attachment for a more detailed breakdown.

Feasibility of Renting Videoconference Rooms

An alternative to building our own facilities would be to rent time in three privately owned videoconference centers. This could be done readily in San Francisco, where there are several centers to choose from. In St. Cloud, MN, however, there are none, but there is an excellent facility at the Technology Center of the University of Minnesota in Minneapolis, 74 miles southeast of St. Cloud. In New Haven, CT, only one videoconference facility is available for rent: at the head office of Multinational Shipping Terminals Inc.

Convenience

Although renting private facilities is the simplest way to introduce videoconferencing into The Courtland Group, there are constraints. The training department estimates it will use the facilities 180 days a year, starting in 2002. (The Board members would require the facilities about 4 days a year.) However, there is no guarantee that a private facility will always be available at the times we need it. This limitation would be exacerbated because all three, or at least two, of the videoconference rooms would need to be available at the same time.

The time difference between San Francisco and St. Cloud (2 hours), and San Francisco and New Haven (3 hours), also introduces a complication, because we will have only a short "window" each day to conduct training, unless one end of the link goes into overtime. A rented facility might not always be able to accommodate such a window.

Cost

The table on page 6 shows the rental rates per day for the three private videoconference rooms, and the total cost based on an annual usage of 180 days a year.

(17) This is a "closed" table, with horizontal and vertical lines separating its compartments.

(18) The **Evaluation** compartment starts here. Leo immediately identifies the criteria against which he will evaluate each method.

(19) Leo needed a *third* level of heading. He chose to incorporate these headings into the text, to clearly differentiate them from the other two levels.

(20) These cost totals provide the base figures for the graph on page 7.

⑰

Location	Rate per day ($)	Cost per year ($)
San Francisco (Oakwood) Distributors	400/day	73,600
Minneapolis (Univ of Minnesota)	450/day	82,800
New Haven (Multinational Shipping)	250/day	46,000
Total per year:		202,400

⑱ **Evaluation of Alternatives**

If we are to introduce videoconferencing, two criteria affect our decision: convenience and cost-effectiveness.

Convenience

An in-house videoconference center offers the most convenience: travel is reduced to a minimum and the room is available whenever we need it. Renting time in a private facility is less convenient because the facility may not always be available and participants have to travel to it (this is a particular inconvenience in Minnesota).

Cost Comparison

To determine whether videoconferencing is fiscally viable, we have projected costs for one year under three conditions: operating as we do now, renting videoconference facilities, and installing our own videoconference centres.

⑲ *Current Method:* The one-year total cost is $800,000, made up of

- Board meetings $150,000
- Training $650,000

Before calculating videoconferencing costs, we have to include the cost of transmitting the video and audio signals via satellite or through fiberoptic data transmission lines. For the usage estimated earlier, we have determined this to be $42,000 annually.

⑳ *Renting:* The one-year cost will be $244,400, made up of

- Rental $202,400
- Transmission $42,000

In-House: The first-year cost will be $265,200, made up of

- Installation $223,200
- Transmission $42,000

In subsequent years the cost will be $42,000 per year.

Plotting these costs on a graph shows that the cost of renting videoconference facilities is the most economical initially (see Figure 3). Over the long term, however, installing our own videoconference centers will become the most economical and most convenient. The break-even point occurs after only 13.3 months.

(21) Although the first-year cost figures on the previous page are clear, they need a simple graph to *show* readers why the initially cost-intensive method is really the best method to choose in the long term. Readers trying to interpret the costs on the previous page may not easily reach the same conclusion. This also satisfies the "visual" person who (22) needs to "see" rather than just "read" the information.

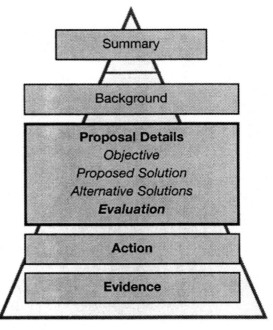

This is the **Outcome/Action** compartment. When making a recommendation, Leo has remembered to use the active voice (he writes: "I recommend...") rather than use the bland, unassertive passive voice (he would have written: "It is recommended that...").

(23) An attachment like this provides the **Evidence** to support statements made in the body of a proposal.

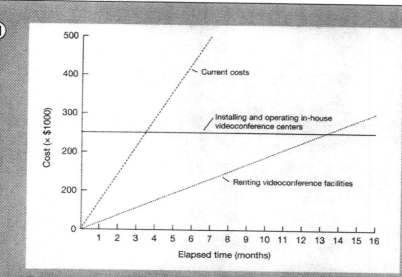

Figure 3. Comparison of videoconferencing costs with current meeting and training methods.

Recommendation   We recommend carrying out an in-depth study into the costs of installing videoconferencing centers in the Courtland Group's offices at San Francisco, CA, St. Cloud, MN, and New Haven, CT.

Leo Cheng

Leo Cheng
Manager, Training and Communication
March 1, 2001

7

Attachment

(23)

Cost Analysis: Equipping Three Videoconference Centers

Service or equipment	Cost per center ($)	Total cost ($)
Renovations		
Physical renovations	17,000	51,000
Wall panelling and dividing wall	2,700	8,100
Floor carpet	3,600	10,800
Lighting and video camera gantry	2,367	7,100
	25,667	77,000
Furniture		
Desks — 12 per center	3,600	10,800
Chairs — 30 per center	6,000	18,000
	9,600	28,800
Equipment		
Video cameras	8,000	24,000
Video monitors	7,200	21,600
Camera control computer	3,000	9,000
Individual positions (mic, switch, etc.)	9,600	28,800
Teaching console (San Francisco)	(7,000)	7,000
	27,800	90,400
Equipment Installation and Testing		
Video cameras, monitors, mics, etc.	8,700	26,100
Teaching console (San Francisco)	(900)	900
	8,700	27,000
Totals: San Francisco, CA	79,666	
St. Cloud, MN	71,767	
New Haven, CT	71,767	223,200

Note: These are conservative cost estimates. Exact costs will be calculated during preparation of a detailed proposal.

Proposals That Offer a Service

The writing compartments for a semiformal proposal that offers a service to a client are shown in Figure 7–3. In this type of proposal the writer

- demonstrates that the proposer understands and is responsive to the client's needs,
- describes specifically what the proposer will do to meet those needs,
- demonstrates that the proposer has the capability to do the work, and
- establishes who (the proposer or the client) will be responsible for implementing specific activities.

A confident writing style and a good-looking proposal appearance also help convince a client that the proposer is the right person or company to hire.

Proposals that offer a service may be either *solicited* or *unsolicited*. A solicited proposal is easier to write, because the audience has been identified and the requirements have been articulated (even if only minimally) by the client. An unsolicited proposal is more difficult to write, because the proposer is offering a service without having first been asked to provide it.

Figure 7–3   Writing plan for a semiformal proposal that offers a service.

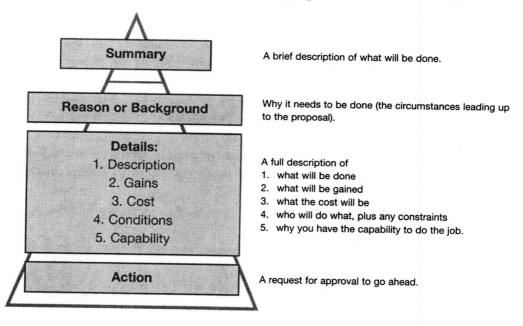

**Semiformal Proposal No. 2
Offering to Provide a
Service**
Comments

Valerie Morrow has written a letter-form solicited proposal. It developed from discussions with a current client (the Western Grain Producers Consortium) and is a stage in an evolving project.

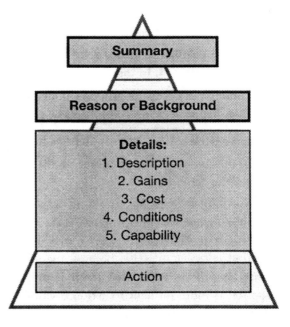

① A letter-form proposal often is more suitable when responding to a request, providing the proposal does not exceed four or five pages. The alternative would be to write a short cover letter and provide the proposal as an attachment.

② The **Summary Statement** describes *briefly* what CommCon Inc. (CCI) is proposing and identifies the total cost. There are two schools of thought about placing the cost right at the start of a proposal. Some people say it's better to identify only what will be done and to present the cost much later, so the reader will not be prejudiced before reading all the details. Others say it's better to include the total cost right up front, because that's the question a reader is most likely to ask after reading the Summary. We subscribe to the latter view. If an organization is known for the quality of its work, a reader will not be "frightened off" by being given the cost up front; rather, the reader will expect the writer to provide a thorough justification in the body of the proposal.

③ The **Reason** or **Background** compartment starts here and continues to the foot of the page. It first describes the sequence of events that led to CCI being invited to prepare a proposal.

④ In addition to describing the circumstances, an introductory section like this should state
  • the purpose of the proposal (described here as "to develop a strategy for training 110 marketing personnel"), and
  • the scope of the proposal (outlined in the five bulleted points and the four numbered objectives).

(1)

CommCon Inc.
PO Box 181, Station C
St. Paul MN 55104
Tel: 651.488.7060
Fax: 651.488.7294

March 6, 2001

Stephen Van Hoight
Director, Human Resources and Training
Western Grain Producers Consortium
3130 South Drive
Minneapolis, MN 55166

Dear Mr. Van Hoight:

(2) I have prepared a proposal for implementing a Certificate Course in Oral and Written Communication for the marketing staff of the Western Grain Producers Consortium (WGPC) in Minneapolis, Minnesota. There will be five training sessions, each comprising a series of five one-day courses spread over 15 weeks between April 30, 2001 and January 25, 2002. The total cost will be $54,000.

Background

(3) The need for the courses was articulated at a January 24 meeting, when Christine Fermore, executive director of WGPC, expressed concern that the communication skills of WGPC's marketing staff have to be upgraded if the Consortium is to meet the increased challenge of selling in international grain markets. CommCon Inc. (CCI) was asked to work with WGPC's Human Resources Department to develop a strategy for training 110 marketing personnel.

(4) At a February 15 focus group we identified several gaps in the performance skills of many WGPC marketing staff. These include the ability to

- write grammatically correct language,
- compose effective letters, memos, and email,
- plan and write convincing, persuasive proposals,
- organize and write effective sales reports, and
- make efficient presentations to clients.

We also established four objectives the training plan must meet:

1. No course must remove participants from their regular work for more than one day at a time (because marketing staff must keep in contact with their clients).
2. The program must develop both written and oral communication skills.
3. The program must be mandatory for all marketing staff.
4. Results must be measurable.

1

⑤ The **Details** compartment and the **Description** subcompartment start here and continue almost to the foot of page 2.

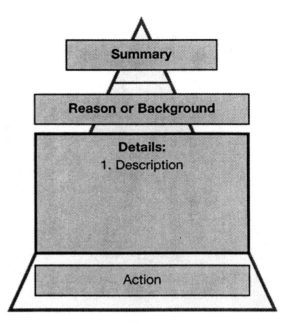

⑥ CCI had already discussed with the WGPC Human Resources department that a standard one-shot course would be inadequate. The reason is re-established here, partly to lead comfortably from the problem to the proposed remedy, and partly because the proposal will be circulated to other managers who may not be aware of the previous discussions.

⑦ The charts showing the schedules are placed in attachments to avoid cluttering the proposal and distracting a reader's attention from the main narrative. However, the narrative must always provide sufficient information so that readers are not tempted to flip through the pages to find the attachments at this point in their reading. This paragraph also demonstrates that no course will be more than one day long, and thus satisfies the first of the objectives established at the foot of page 1.

⑧ Headings are essential in a long letter-form proposal like this, to guide the reader, to create a more attractive appearance, and to break up long stretches of narrative.

⑨ The five descriptions could have been simply summarized as one-line statements here, and the details placed in an attachment. The descriptions have been kept here because they satisfy the five deficiencies listed in the bulleted points in the Background (see page 1 of the proposal).

⑩ It's essential that the five courses are listed in *exactly the same sequence* as the deficiencies listed on the previous page. To describe them in a different sequence would confuse the proposal's readers.

⑪ At this point the list satisfies the second objective established on page 1 (that both written and oral communication skills be taught).

The Plan

(5)

(6)

The difficulty with most courses designed to improve a person's communication skills is that the instruction and the experience tend to be transitory. Participants finish the course feeling they have learned some useful skills, yet gradually they slide back into their old habits and the value of the training is lost or at best diminished. This is particularly true of writing courses.

We propose to stretch each person's training over a 15-week period, hold a review session approximately every four weeks, and measure progress regularly. Each participant will attend four one-day written communication modules, as illustrated in Attachment 1.

(7)

Selected participants will also attend an additional one-day module on oral communication techniques. A suggested schedule for the courses is shown in Attachment 2. The five modules will cover the following topics:

(8)

- *Module 1: Language Handling (Week 2)*

  Participants will learn how to construct effective sentences and paragraphs, write clearly and concisely, and avoid common writing pitfalls. They will then apply the techniques to letters, memos, and email.

- *Module 2: Review and Interviews (Week 6)*

  One week prior to module 2, participants will submit three pages of writing they have done during the past two weeks. The instructors will evaluate their work and then plan a two-hour session that reviews the common errors found in the examples. For the remainder of the day, the instructors will meet course participants individually to advise and coach them.

(9)

- *Module 3: Proposal Writing (Week 9)*

  Participants will learn how to plan and write convincing marketing proposals. They will be asked to bring an "in process" proposal with them and work on it during the session, with guidance from the instructors.

(10)

- *Module 4: Report Writing (Week 12)*

  Participants will learn how to organize and write informative sales and progress reports. They will be asked to bring an "in process" report with them and work on it during the session, with guidance from the instructors.

(11)

- *Module 5: Oral Presentations (Week 15)*

  Participants will learn how to prepare for and make effective sales presentations, which will be videotaped for their future review. They will also learn how to prepare computer-generated visual aids, and display them with an LCD projection system.

(12) This paragraph satisfies the third objective on page 1 (that the program must be mandatory) and simultaneously defines the style of teaching to be used.

(13) By identifying how the course participants' performance will be measured, this section fulfills the fourth objective on page 1. Although not stated in the proposal, *measuring* a person's capability in written or oral communication is difficult to achieve. The proposal anticipates that Stephen Van Hoight—its primary reader—will know this, and so describes in some detail how it will be done.

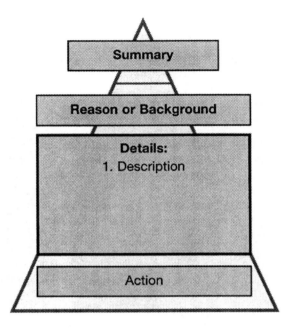

(14) Here, Valerie describes the company's philosophy toward teaching a "soft skill" course such as writing. CCI knows its approach is unique and so identifies the advantages that will accrue if the courses are taught using its techniques. This section is a continuation of the **Description** subcompartment.

(12) Modules 1 through 4 will be mandatory for all marketing staff, so the concepts will be used consistently throughout WGPC. Module 5 will be optional. The emphasis in modules 3, 4, and 5 will be on providing practical hands-on experience for participants. Each module will start with a review of previous instruction.

(13) Evaluating Performance

The participants' progress will be evaluated in three ways:

1. By the **course instructors**, who will assess individual writing performance by having participants complete writing exercises during each module.

2. By each **participant's manager**, who will complete a unique evaluation form developed by CCI. The assessment will yield an objective "score" and will be administered three times:

   • In week 1, before the participants receive any training.

   • In week 8, when the participants have completed modules 1 and 2.

   • In week 14, when the participants have completed all four writing modules.

3. By each **participant**, who will complete a personal evaluation form developed by CCI. This form also will yield a "score" and will be administered twice:

   • In week 2, at the start of module 1.

   • In week 14, following completion of all four writing modules.

Only the course instructors will evaluate participants' oral presentations.

CCI's Methodology

(14) We believe in presenting highly interactive courses, with participants having plenty of hands-on practice in applying the concepts we teach, and maximum opportunity for personalized instruction to correct individual behaviors. To achieve this, we consistently teach with two instructors.

We are aware that some participants come to our classes with trepidation, having had an unhappy experience with high school language arts classes or post-secondary education in writing and speech. Consequently we present our ideas in an easy-to-comprehend-and-apply format, which speeds and facilitates learning for all participants. By teaching the unique "pyramid" method for organizing a letter or report in the first hour, we demonstrate a direct way to start writing they can adapt readily to their regular work. During the remaining hours and future courses, they practice applying the pyramid to various types of letters, reports, and proposals.

⑮ Now that the reader understands CCI's teaching philosophy, Valerie introduces the advantages that will accrue to WGPC if the proposal is accepted and implemented in its present form. This becomes the **Gains** subcompartment.

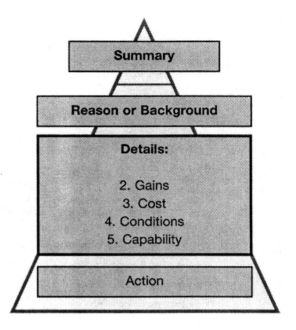

⑯ Describing **Cost** is always difficult. The proposing company needs to present sufficient details so that the reader understands how the total cost has been developed, yet not swamp the reader with extensive details. If a client asks for a full description, the costs should be presented as an attachment with only the key figures shown in the body of the proposal. Here— and particularly because CCI's pricing for other projects is already known to WGPC—only totals are presented in the proposal.

⑰ Having identified the cost structure, Valerie immediately states the assumptions on which the cost calculations are based. These are also the **Conditions** that will apply to the project. They fall into two categories: what service the proposing company will provide, and what aspects the client will be responsible for.

It's essential to identify such factors in a proposal, otherwise partway through the project a disagreement may occur because each party has *assumed* the other party will be responsible for certain aspects. Other Conditions in a proposal to provide a service might define the need for progress payments during the life of the project, or a clause that identifies the fee to be paid if the project is canceled partway through.

⑱ Although CCI and WGPC have worked together on the initial development of the project, and CCI has provided previous training for the Consortium, the proposal still has a **Capability** section. This recognizes that some decision-makers at WGPC may be less aware of CCI's experience and competence. It's also possible that the client may seek competitive proposals, each of which will describe the proposing company's capabilities, against which CCI's capabilities will be compared.

Participants are also asked to bring in samples of their writing throughout the modules, so we can provide immediate feedback.

(15)   **Advantages**

By continually measuring performance, the course instructors will be able to determine each participant's evolving skills as the course progresses, and identify participants who need additional advice. It will also demonstrate to participants that WGPC has undertaken an in-depth program, is monitoring progress, and is recognizing quality achievement.

Because participants will be attending courses regularly over 3 1/2 months, they will be continually reminded of the techniques they have been taught, and will have guided practice in applying them. They will be much less likely to regress into old habits.

(16)   **Cost**

CCI's fee for developing, facilitating, and evaluating the participants' work will be calculated on a teaching-day basis. For each teaching day of the written communication courses the fee will be $2200 for 22 course participants. For each teaching day of the oral communication course the fee will be $2000 for 12 participants (assuming that 60 of the 110 marketing staff will attend the oral communication course). This calculates out as:

- 5 written communication courses for 4 days each @ $2200 per day: $44,000
- 5 oral communication courses for 1 day each @ $2000 per day: $10,000

Total Cost: $54,000

(17)   These costs are based on CCI providing two instructors for each course, LCD projection equipment, instructional videos, handout originals for WGPC to duplicate, and a report of each participant's capabilities and improvement.

Our proposal also assumes that the courses will be held on WGPC's premises, that WGPC will provide a screen and flip chart, and that WGPC will duplicate the requisite number of handout copies for each course.

CCI's Experience in Developing and Facilitating Communication Courses

(18)   CCI has been providing courses in technical and business communication for over 20 years, in the US, Canada, and Europe. Our presentations range from correspondence courses for people living in remote locations, through in-house courses for targeted groups of employees, to open registration courses in major centers. A representative list of clients is shown in Attachment 3.

4

⑲ Because of the ongoing relationship between client and proposer, this is a very direct **Action Statement**. CCI *expects* to be drawn into further discussions and to carry out the work if WGPC decides to go ahead with the project.

⑳ There are two **Attachments**, both of which demonstrate that the proposer has developed a detailed plan for implementing the training. (A third Attachment—containing the proposing company's client list—is not included here.) It would have been too early, and too costly, to develop detailed training plans and course outlines for a project that may not go ahead, or may go ahead in an altered form. They will be prepared after the client indicates readiness to go ahead with the project. This would be the "next step" mentioned in the **Action Statement**.

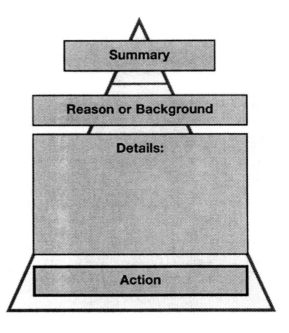

In the US Mid-West, we have provided courses for numerous power and telephone utilities, for Multiple Industries, and for the Western Grain Producers Consortium. For WGPC, we have provided in-house workshops for managers (1999–2001) and distance-education courses for supervisors at remote sites (1994–2000).

Every course we teach is tailored to suit the specific group of employees we meet or correspond with. This tailoring starts during initial discussions with a client, when courses are being planned, and continues during the course when we meet the participants and identify their individual learning needs.

(19) We appreciated the help provided by Kim Andersen, of your Human Resources Department, throughout the preparation of this proposal. Please call me if you have questions, and let me know when you are ready to schedule a meeting to plan the next step in the training program.

Sincerely,

Valerie Morrow

Valerie Morrow
Project Coordinator

att: 2

Attachment 1

Proposed Teaching, Review, and Progress Evaluation Plan

Week No.	Module No.	Primary Topic	Evaluation Method
1			Supervisors assess writing competence
2	1	Language Handling	Instructors evaluate writing competence
3			
4			
5			
6	2	Review + Interviews	Instructors evaluate writing competence
7			
8			Supervisors assess writing competence
9	3	Proposal Writing	Instructors evaluate writing competence
10			
11			
12	4	Report Writing	Instructors evaluate writing competence
13			
14			Supervisors assess writing competence
15	5	Oral Presentations	Instructors evaluate speaking competence

Notes:

1. Each module lasts one day.

2. The written communication course comprises four modules = four days.

3. The oral communication course is one module = one day.

4. We estimate 60 of the 110 participants will enroll in the oral communication course.

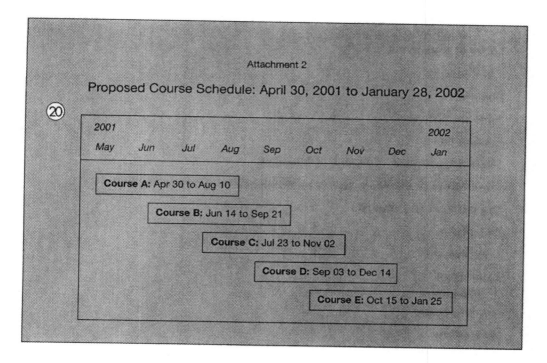

Attachment 2

Proposed Course Schedule: April 30, 2001 to January 28, 2002

20

2001								2002
May	Jun	Jul	Aug	Sep	Oct	Nov	Dec	Jan

Course A: Apr 30 to Aug 10

Course B: Jun 14 to Sep 21

Course C: Jul 23 to Nov 02

Course D: Sep 03 to Dec 14

Course E: Oct 15 to Jan 25

The Formal Proposal

Formal proposals are normally lengthy documents that sometimes run to several volumes. Hence, their size prohibits a sample from being included here. Instead, a typical outline and the purpose of each compartment are described below.

Most formal proposals are written in response to a Request for Proposal (RFP) issued by the government or a large commercial organization. Normally time is short between the date an RFP is issued and the date the proposal must be presented to the originating agency. Companies submitting proposals each form a proposal team of key individuals, who work long hours to ensure their proposal is written, illustrated, printed, and delivered before the closing date.

Many agencies issuing RFPs stipulate the major topics the proposing company must address and the sequence in which information is to be presented. Unfortunately, although there is some similarity between the formats stipulated by different agencies, there are sufficient variances to make it impossible to present a standard outline here.

The major components of a proposal are illustrated in Figure 7–4. When the compartments are converted to headings they form an outline, such as the outline illustrated on the following page, which is a simplified composite of several outlines and is applicable to either a solicited or an unsolicited (i.e. company-initiated) proposal. The primary information-bearing parts are shown in boldface type:

Cover

Letter of Transmittal

Title Page

Summary

Table of Contents

Introduction

Description of Work, Problem, or Situation

Approach to Doing Work, Resolving Problem, or Improving Situation

Organization and Planning

Exceptions

Price Proposal

Experience:
   Company
   Employees

Appendixes

Figure 7–4   Typical writing plan for a formal proposal.

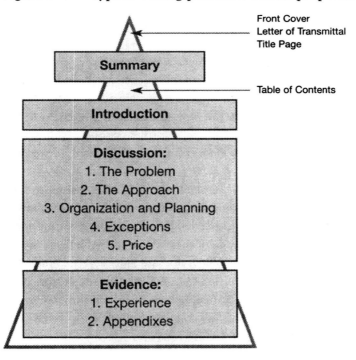

Letter of Transmittal

When attached to a formal proposal, a letter of transmittal assumes much greater importance than the standard cover letter attached to the front of a semiformal or formal report. Normally signed by an executive of the proposing company, it comments on the most significant aspects of the proposal and sometimes the cost. As such it has a role similar to the executive summary that sometimes precedes a formal report (see page 125).

Summary

The summary mentions the purpose of the proposal, touches briefly on its highlights, and states the total cost. If an executive summary or letter of transmittal is bound inside the report, the summary is sometimes omitted.

Introduction

The introduction section of a proposal is like the introduction in a report: it describes the background, purpose, and scope of the proposal. If the proposal is prepared in response to an RFP, reference is made to the RFP and the specific terms of reference or requirements imposed by the originating authority.

Description of Work, Problem, or Situation

This section describes the work that needs to be done, the problem that has to be resolved, or the situation that needs to be improved. It usually includes

- a statement of the work/problem/situation, as defined by the RFP,
- an elaboration of the work/problem/situation and its implications (to demonstrate the proposer's full comprehension of the circumstances), and
- the proposer's understanding of any constraints or special requirements.

Approach to Doing Work, Resolving Problem, or Improving Situation

The proposer describes the company's approach to the work/problem/situation, states specifically what will be done and why it will be done, and then in broad terms outlines how it will be done. This section must be written in strong, definite, convincing terms that will give the reader confidence that the proposing company knows how to tackle the job.

Organization and Planning

Here, the "how" is expanded to show specifically what steps the proposer will take. Under Organization the proposer describes how a project group will be established, its composition, its relationship to other components of the company, and how it will inter-act with the client's organization. Under Planning the proposer outlines a complete project plan and, for each stage or aspect, describes exactly what steps will be taken and what will be achieved or accomplished.

Exceptions

Sometimes a company may conceive an unusual approach that offers significant advan-tages, yet deviates from one or more of the client's specified requirements. Here, the proposing company lists the exceptions, explains why each requirement need not be met, and describes the advantages to be gained by taking an alternative approach.

Price Proposal

The proposer's price for the project is stated as an overall figure and broken down into schedules for each phase of the project. The extent and method of pricing is often spec-ified by the RFP.

This section of the proposal is the one most likely to be found in varying positions. The RFP may stipulate that it appear here, at the front, as the last section, or even as a separate document.

Experience

The proposing company describes its overall experience and history, and its particular experience in doing work, resolving problems, or handling situations similar to those described in the RFP. Key people who would be assigned to the project are named, and their experience is described in a resume (sometimes referred to as a *curriculum vitae*, or CV).

Appendixes

The appendixes contain supporting documents, specifications, large drawings and flow-charts, schedules, equipment lists, etc, all of which are referenced in the proposal.

Proposal Appearance

Major proposals are multipage documents assembled into book form. They are usually bound with a multiring plastic binding or the "Perfect" binding method used for soft-cover books. Minor proposals have fewer pages but are still bound or stapled into book form. Some very short proposals, particularly those submitted from one company to another, may be simply stapled together like a semiformal report or even, in some cases, be submitted as a letter.

PART 4

Formal Reports

CHAPTER 8

The Formal Report

Formal reports have a much more commanding presence than informal or even semiformal reports. Usually bound within a simple but dignified jacket, they immediately create the appearance of an important document. Internally, their information is compartmented and carefully spaced to convey confidence from start to finish.

The term *formal report* refers to the type of document rather than its title. A formal report is more likely to be referred to as a feasibility study, an investigation or evaluation report, a product analysis, or a project report. Sometimes one of these names may precede the report's main title, but more often the name is omitted and the title stands alone (for example, the words Evaluation Report or Investigation Report do not appear on the title page shown on page 129).

The following sections show two ways to organize a formal report: the traditional approach and an alternative approach.

Traditional Arrangement of Report Parts

There are six main compartments in a formal report (see Figure 8–1):

Summary

Introduction

Discussion

Conclusions

Recommendations

Appendix

Figure 8–1   Writing plan for a traditional formal report.

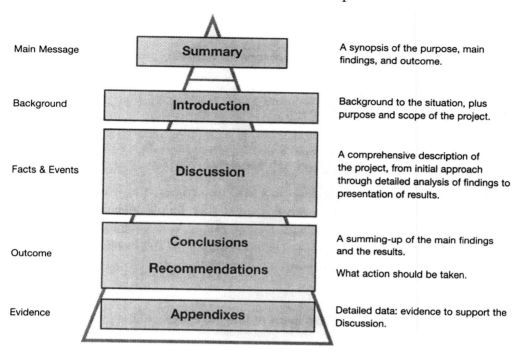

The first four compartments are identical to the four basic compartments identified in Chapter 2. If the writer of a formal report also proposes that action be taken, then the report has a Recommendations compartment. And if it contains supporting data, it has an Appendix.

The six main report parts identified in Figure 8–1 are the primary information-bearing compartments of a formal report. There are also additional parts that support these main compartments, and they help give the report its formal shape. They are listed in their appropriate positions in relation to the main compartments:

Cover Letter

Cover Page

Title Page

Summary

Table of Contents Page

Introduction

Discussion

Conclusions

Recommendations

References or Bibliography

Appendix

Back Cover

Guidelines for writing these parts appear on the following pages, facing the relevant sections of Dan Rogerson's report remedying the decrease in sales at Provo Department Stores' catalog order centers.

Alternative Arrangement of Report Parts

To meet the needs of executive readers, sometimes the Conclusions and Recommendations are brought forward so that they follow immediately after the Introduction. This rearrangement helps readers gain a more complete picture of the report's outcome without having to read all the details of the Discussion. Coincidentally, it forms the three-tiered pyramid shown in Figure 8–2.

Figure 8–2   Alternative arrangement of writing compartments for a formal report.

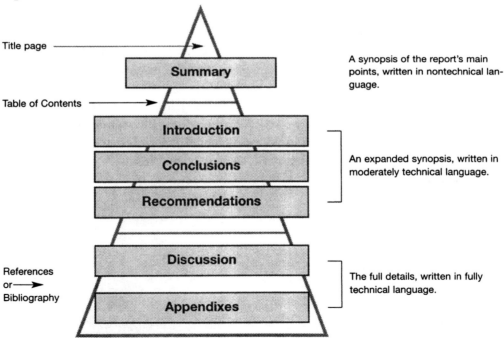

Moving the Conclusions and Recommendations forward does not materially affect the report's content, but it does demand careful attention by report writers. They must ensure that

- the transition from Introduction to Conclusions is achieved smoothly, naturally, and coherently,

- the Conclusions and Recommendations can be clearly understood (since their readers will not yet have read the Discussion), and

- the Discussion does not start and end too abruptly. Since it is no longer preceded by the Introduction or followed by the Conclusions, it will probably need a new introductory paragraph and an additional closing paragraph.

Analysis of a Formal Report

The following pages contain

- the cover letter accompanying a report prepared by Dan Rogerson of L. V. Morton and Associates (Figure 8–3),

- the report itself, on odd-numbered pages 129 to 153, and

- comments on how each part of a formal report should be written (on even-numbered pages 128 to 150), with some comments on how Dan wrote his report.

Cover Letter

A cover letter is really a transmittal document used to convey a report from one organization to another (or one person to another). The letter is paper-clipped to the *outside* of the report's front cover (in effect, the report is attached to the letter) and is not part of the report.

A cover letter can be a simple statement:

I enclose our report "Strategy for Remedying the Decrease in Sales at Provo Catalog Order Centers," which has been prepared in response to your letter of January 12, 2002, and our discussions on December 18, 2001.

Alternatively, it can offer comments on specific aspects of the report that will be of particular interest to company executives, as Dan Rogerson's cover letter does. A detailed cover letter like this is known as an **Executive Summary**, because it contains information directed specifically to corporate executives. Such information may deal with special factors or political or financial considerations that would be either unsuitable for every reader or out of place in the report itself. The executive to whom the cover letter is addressed normally removes the executive summary before circulating the report to other readers.

You may occasionally encounter a second type of executive summary that is bound within the report itself immediately after the title page. Normally it will contain the title **Executive Summary** centered at the top of the first page, which will be followed by one to three pages (depending on the length of the report) that summarize the key elements of the full report. Because it is bound *within* the report, this type of executive summary does not comment on special factors or political or financial considerations that might not be suitable for every reader.

Figure 8–3   A cover letter that also serves as an executive summary.

L. V. Morton and Associates

Business and Computer Consultants
105A George Street
Rochester NY 14607

February 28, 2002

Marianne Lamoureux
Director of Marketing
Catalog Order Division
Provo Department Stores
2320 Utica Boulevard
Syracuse NY 14471

Dear Marianne:

I have researched methods for increasing customer acceptance of online ordering, as requested in your letter of January 12, 2002, and am recommending installing video cameras in the Provo catalog order centers, with links to your customer service representatives in Rochester, NY. The cost will fall comfortably within your budget for years 2002 through 2006.

As we discussed early in the project, the simplest and most economical method would have been to re-install telephone handsets and use the existing lines between the order centers and Rochester. Unfortunately, for technical reasons it would not be possible to run both the telephone messages and the computer signals on these older lines, which now are being used solely for computer use. To install new fiber-optic lines at this stage, which could handle numerous messages, would have considerably exceeded the budget.

The discounted equipment prices I have negotiated with the suppliers are good until May 31, 2002.

Thank you for inviting us to research methods for you. If you have questions, please call me at (716) 447 1691.

Sincerely,

Dan Rogerson

Dan Rogerson
Senior Consultant
L. V. Morton and Associates

enc: report (3 copies)

Title Page

The title page contains four main elements:

- The full title of the report, which must be informative without being too long. Dan's title, though a little long, is informative. A brief title such as **Problem at Catalog Order Centers** would not have been sufficiently definitive.

- The name of the organization and sometimes the person for whom the report has been prepared.

- The name of the originating organization and sometimes the name of the person who has written the report.

- The date the report is issued. If a report number is also included on the title page, it and the date are offset left and right. If there is no report number, the date is moved to the center, as shown.

A title page must have visual appeal, yet be simple and dignified. Every line must be centered individually on either side of a vertical centerline, which is offset about 0.25 in. to the right of page center to allow for the unusable 0.5 to 0.75 in. on the left-hand edge of each page (see Figure 8–4). This edge is usually covered by the binding or is punched for a multiple-ring binder. If a report is printed on both sides of the paper, the binding edge on the reverse side of each sheet is on the right-hand edge of the page, and the centerline is offset about 0.25 in. to the left. Most word-processing programs make these adjustments automatically when you define your printing requirements.

Figure 8–4   The typing centerline is offset from the page centerline.

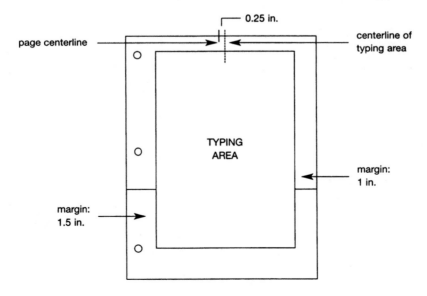

Strategy for Remedying the Decrease in Sales at Provo Catalog Order Centers

prepared for

Marianne Lamoureux

Director of Marketing

Catalog Order Division

Provo Department Stores

Syracuse, New York

prepared by

Dan Rogerson

Senior Consultant

L. V. Morton and Associates

Rochester, New York

February 28, 2002

Summary

In a formal report the Summary immediately follows the title page but appears before the Table of Contents. It always has a page to itself and is centered on the page. Sometimes it is preceded by the heading **Abstract** instead of **Summary**.

The Summary is the most important page in the report. Because it is the first body of information that readers encounter, it has to encourage them to read further; if it does not, it has failed to achieve its purpose.

Guidelines for writing summaries are:

- Write the Summary after the rest of the report has been written, but place it at the front of the report.

- Indent the Summary from both left and right margins, and center it vertically on the page.

- Draw information for the Summary from the Introduction (particularly the purpose of the project), the Discussion (pick out the most important highlights), and the Conclusions and Recommendations (the outcome or result of the project).

- Keep the Summary as short as possible and make it interesting and informative. For example, rather than write "conclusions are drawn and a recommendation is made," state specifically what the main conclusions are and what you are recommending or suggesting should be done.

- Keep the intended readers clearly in mind to ensure you tell them what they most want to know or need to hear.

- Use plain, nontechnical words and avoid topic jargon, so that the summary can be understood by almost any person who reads the report.

Summary

When, in the summer of 2000, Provo Department Stores replaced the telephone ordering equipment with a computer ordering system at its mall-based catalog order centers, the company experienced an immediate 40% decrease in ordering activity. The decrease remained constant over the following six months. The problem was traced to over 50% of catalog order customers who, having little or no experience in working with computers, felt uncomfortable trying to place their orders and missed the human contact they experienced when ordering by telephone.

The remedy is to reintroduce the human element into the ordering sequence. By gradually educating customers to order online, Provo will increase customer confidence in using the computers in the catalog order centers.

The most effective method will be to install a video camera in each catalog order center, and also at the customer-service representatives' desks, so that customers needing help will be able to see *and hear* the customer-service representative who is giving them advice. The cost will be $11,173 less than the remedial budget established by Provo for 2002, and $7647 per year less than the budget established for subsequent years.

i

Table of Contents

A table of contents (T of C) is inserted at the front of a report mainly to help readers find specific information. But it also has a secondary, much more subtle purpose: to let readers *see* how the author has organized the information and what topics are covered.

Readers who have read no more than Dan Rogerson's Summary can quickly establish his approach from the T of C. They will see the logic of his organization and how the primary headings (those between the Introduction and the Conclusions) will

- discuss the alternatives for increasing customer acceptance of online ordering,

- describe the various kinds of equipment and how they would be set up, and

- evaluate which equipment and setup is most suitable.

These are the factors you should take into account when creating a T of C:

- Every major topic heading in the report must also appear in the T of C.

- The topic headings in the T of C must be worded exactly as they appear in the report. This can be accomplished readily if you use word-processing software to create the T of C for you.

- Minor subordinate headings may be omitted from the T of C if their inclusion would make the T of C too lengthy or detract from the clarity of the overall organization plan.

- The page heading often is more appropriately shortened to the single word **Contents**, as has been done here.

- All appendixes must be listed, with the complete title drawn from the first page of each appendix.

- If drawings or illustrations are grouped separately in the report, they should be listed in the T of C. If there are many illustrations, it is acceptable to insert the single entry "Illustrations" and page number in the T of C and to place a separate list of illustrations as the first page of the Illustrations section.

Contents

Appendixes

Introduction

(1) The Introduction prepares readers for the details that follow in the Discussion. It introduces them to the circumstances leading up to the project, and to the reasons it was undertaken and the report was written.

An Introduction has three main components:

- The **Background**, which describes events leading up to the existing situation, what projects (if any) have been done previously, and why the current project or study is necessary.

- The **Purpose**, which defines what the project or study is to achieve, who authorized it, and the specific terms of reference.

- The **Scope**, which outlines any limitations imposed on the project, either by the person(s) authorizing it or by the person undertaking it, such as cost, time in which the project is to be completed, depth of study, and factors that must be included or may be omitted.

You do not have to present these components in this order, although Dan Rogerson happens to have done so in the Introduction to his report. Paragraphs 1 to 4 plus the three bullets are the **Background,** and the last paragraph contains both the **Purpose** and the **Scope.**

In very long reports, the three components can be treated as separate topics preceded by individual headings.

(2) This is the first of six references Dan makes to information he has gleaned from other documents. He lists the documents in full at the end of his report (see report page 8 and comments (6), (7), and (25)).

(3) The **Purpose** and **Scope** are identified clearly here. The wording Dan used for this paragraph is drawn almost directly from the client's letter of Authorization.

Strategy for Remedying the Decrease in Sales
at Provo Catalog Order Centers

Introduction

① In addition to its over-the-counter retail marketing, Provo Department Stores (Provo) has operated a telephone-based catalog order system since 1956. Initially, customers telephoned orders in from their homes, but in 1978 Provo set up telephone-equipped catalog order centers in its department stores. Two years later Provo also installed telephone catalog order centers in malls where Provo did not have a presence.

Each order center contained a 6 ft counter with catalogs placed on it, two chairs, and two direct-line telephones connected to the order center in Provo's Warehouse at 2200 Ridgeway Avenue in Rochester, New York. By 1991 there were 590 booths, all unstaffed.

By the mid-1990s, however, Provo recognized that its catalog order business would have to go online, and so in 1997 the company started accepting orders over the Internet from its home-based customers. Because not all of Provo's catalog order customers would have a home computer or access to the Internet, the company also recognized that it would have to continue operating its telephone ordering system in parallel, at least until 2006.

In early 2000 Provo converted 564 of its stand-alone catalog order centers from telephone order stations to computer order stations, and in each installed two video monitors, CPUs, keyboards, and a large-print backboard containing detailed instructions for ordering online. However, over the next six months orders from the catalog order centers decreased by over 40%. To determine the cause, Provo engaged the Dolman Survey Group (DSG) to identify the public's reaction to the

② change. DSG's report[1] shows that customers who experienced difficulty placing orders

- were unaccustomed to using the Internet, and so chose not to,

- became frustrated when the system "didn't do what they expected," and so simply walked away from the booth without completing their order, and/or

- missed the human element, i.e. speaking to an operator who could answer questions and direct them.

③ In a letter dated January 12, 2002, Marianne Lamoureux, Provo's Director of Marketing, authorized L. V. Morton and Associates to investigate methods that would reverse the downward trend and increase customer acceptance of online ordering. Provo established a budget of $75,000 for year 2002, and $20,000 per year for years 2003 through 2006.

Discussion

④ The Discussion starts here. Note that

- the word *Discussion* seldom appears as part of a heading and is never used as a single-word heading, and

- the Discussion may follow immediately after the Introduction (i.e. on the same page), or it may start on a fresh page as has occurred here.

⑤ How you arrange the information within the Discussion is extremely important. The overall logic of the case you present must be clear to readers, otherwise they may follow your line of reasoning with some doubt or hesitation. This, in turn, may affect their acceptance or rejection of your conclusions and recommendations.

Three factors can have a particularly negative effect on readers:

- Writing that is beyond their comprehension; that is, uses technical terms and jargon they may not understand.

- Writing that fails to answer their questions or satisfy their curiosity; that is, does not anticipate their reactions to the facts, events, and concepts you present.

- Writing that either underestimates or overestimates the readers' knowledge; that is, assumes they know more (or less) about the topic than they really do.

These common pitfalls can be avoided if you clearly identify your readers. You have to establish first whether your report will be read primarily by management, by specialists knowledgeable in your field, or by lay people with very limited knowledge of your specialty. Then you must decide what they are most interested in and what they need to hear from you (particularly if there is likely to be a conflict between what they would like to hear and what you need to say).

Finally, you have to plan your report so that the order in which you present information will answer not only the readers' immediate questions, but also any questions they may generate while they read. For a start, go back to the terms of reference you were given by the person or organization authorizing your project. Pick out the points of most interest to your reader(s), jot them down, and then rearrange them into a logical sequence that will satisfy the readers' curiosity in descending order of importance.

⑥ Readers not only want to be given facts, but they also want to know how they are derived. If it is not practicable to provide an in-depth discussion of certain facts (possibly because they would be irrelevant to the report's main thrust or divert readers' attention), then refer to the source from which the information was obtained. Place the references at the end of the report and call them "References."

(4) ## Increasing Customer Acceptance of Online Ordering

The Dolman Survey Group's research shows that reintroducing the human element into the catalog ordering process at Provo order stations will be a key factor in reasserting customer confidence.[2] We have examined the demographics of the Provo customers interviewed by DSG, and have identified the majority of customers as middle-aged or seniors, who in general are less familiar and less comfortable with computer technology than their more youthful peers. However, simply reverting to telephone ordering from the order stations would be a backward step and would not be a viable answer. (Provo recognizes that customers who order directly from their homes will continue to place their orders by telephone for at least five more years. For customers who do have home computers, Provo is offering free software to encourage them to place their orders online.)

(5) Consequently, Provo needs to accommodate two sets of customers at its catalog order stations: those who are already placing their orders online, and those who are hesitant to try or experience difficulty with computer ordering. By helping the latter group use the keyboard and order software, Provo can gradually convert them to ordering online. Research shows that, once older people become accustomed to using the keyboard, the Internet, and the online ordering program, most become enthusiastic users and proud of their new-found capability.[3] Long-range, this should

(6) re-establish these customers as dedicated, regular catalog buyers.

(7) To re-establish human contact for its order center customers, Provo will need to install either backup telephones (for voice communication) or video cameras (for visual and voice communication):

Alternative 1: Voice Communication

This is the medium customers were accustomed to using before computer keyboards and video monitors were installed in the catalog order centers. For solely voice communication, each individual order station will require a telephone handset and a dedicated telephone line to a Provo customer service representative in Rochester.

Alternative 2: Video Communication

This would be a totally new means of communication between customers and the Provo customer service representatives. A very small video camera would be mounted above the customer's video monitor, with a second video camera mounted above the service representative's screen. A customer needing help would move the cursor to a "Help" button and click on it, and the help call would be routed to an available service representative. On a corner of the video screen, the viewer would see a head-and-shoulders view of the customer service representative; similarly, the service representative would see a head-and-shoulders view of the customer. (If all service representatives are talking to other customers, a "Please Wait" message would appear on the customer's video monitor, much as an automated telephone response system asks a caller to wait for the next available representative.)

The raised 3 in line 7 of this paragraph refers readers to the third item listed in the References on report page 8. (*Note*: Footnotes, which once were popular, are not used in modern reports, partly because they distract a reader's eye but mostly because making room for them at the foot of a page creates layout problems).

(7) This short paragraph is known as an "overview statement." By introducing a series of short paragraphs or subparagraphs, it prepares readers to *expect* alternatives to be presented to them. There is a second example on report page 5, immediately after the heading "Evaluation Plan."

(8) Dan uses a two-step approach to lead his readers into the technical details. At the foot of page 2 he introduces the alternatives and, because the second alternative is a relatively new, untried concept, on this page he describes in general terms how it will work. This is essential if readers are to accept the technical details that will follow.

(9) Although Marianne Lamoureux (the director at Provo Department Stores to whom the report is addressed) and Dan are on a first-name basis, he chooses not to use the first-person "I" in his report because he knows the report will go to other executives for financial approval. By using "we" (as in "we anticipate") he is able to maintain the active voice. Note, however, that he does use "I" in his cover letter, recognizing that Marianne will remove the letter before circulating the report.

(10) Here, Dan is preparing the reader to *expect* that he will consider four possible configurations, even though initially his report seems to be evaluating only telephone vs video methods of communication. This will help the reader to adjust to the four-options evaluation he presents on page 5 of the report (at point 17). Signalling one's intentions is important when steering a reader through a fairly complex report.

(11) Different sizes of **bold typeface** help readers automatically assess which are primary and which are secondary headings (and, concurrently, how the different sections and paragraphs are subordinated).

(12) Extensive details, such as manufacturers' catalogs, tabulated data, calculations, specifications, and large drawings, are placed in an appendix and stored at the back of the report where they will not interfere with reading continuity. They constitute evidence that supports and amplifies what is said in the body of the report.

When selecting and referring to appendixes, ensure that

- they are necessary and relevant,
- every appendix is referred to in the report narrative,
- readers do not have to refer to the appendixes to understand the report.

To prevent readers from having to flip pages and refer to appendixes as they read, a report writer may have to include a synopsis of an appendix's highlights in the report narrative or draw a conclusion from the appended data. For example:

⑧ The customer and the service representative can now talk to each other, their voices being transmitted through the microphones and loudspeakers beside or built into their respective video monitors. The representative answers the customer's questions and can show visually, by moving the image on the screen, which button is to be engaged or highlighted, or where information is to be typed in.

A dedicated telephone line would not be required, because the video and voice signals would be carried on the computer line between the order station and the customer service representative.

The Setup for Customer Service Representatives

⑨ For either method, Provo will have to provide additional equipment at its customer service "Help" stations in its Rochester office. Initially, we estimate up to 15 augmented Help stations will be required at peak ordering periods. After six months, however, we anticipate only between 8 and 10 augmented Help stations will be needed. By then customers will have become accustomed to ordering by computer and the number of help calls will drop at least 30% and possibly 40%.[4]

Each customer service representative will require

- additional software for the video monitor, to show the screen the customer is working on, *and*
- a telephone handset, for solely voice communication, *or*
- a video camera, mounted above the video monitor, for combined video and audio communication.

The Setup at Provo Catalog Order Centers

There are two order stations at each catalog order center, which offers four possible configurations for the proposed communication equipment:

1. One telephone handset and dedicated telephone line, to be shared by the two computer workstations.

⑩ 2. Two telephones and two dedicated telephone lines, one for each computer workstation.

3. One video camera between and slightly above the two workstations, programmed to turn to the customer requesting help.

4. Two video cameras, one installed above each computer workstation.

Equipment Selection

Telephone Handsets

⑪ The most suitable, robust, and economical telephone handset is the Tele-Spirit 300 manufactured by Tele-Center Industries of Philadelphia, Pennsylvania. For a minimum purchase of 500 handsets, the price would be $9.85 per unit; for 1000 handsets, the price will drop to $9.25 per unit.

> A survey conducted in the Camrose shopping center (see Appendix K) showed that during weekdays 74% of shoppers traveled by private automobile and 23% traveled by bus. On Saturdays the number of automobile travelers increased to 83%, and bus travelers decreased to 15%. The remaining 2–3% traveled by bicycle or on foot.

(Appendix K contained seven pages of numerical data; if placed in the report, they would have both physically and psychologically interrupted the narrative.)

⑬ Dan uses two kinds of tables in his report. This is an "open" table, which means it has no border surrounding it or lines separating the groups of figures. The table at ⑭ on report page 5 is a "closed" table, because it has a ruled border and lines separating the main compartments.

⑭ This table summarizes the total cost for each configuration. Because how the costs are derived is fairly complex, Dan places the detailed calculations in an appendix so they will not interrupt reading continuity, providing just sufficient details here for readers to follow his discussion.

⑮ This is the start of a **comparative analysis,** or justification, in which Dan compares the telephone alternative against the video alternative. The key to writing an effective comparative analysis is to *avoid comparing one alternative directly with another*. This may sound like an oxymoron, but in practice it works extremely well.

A comparative analysis contains three main parts which

- describe the items to be compared (the alternatives),
- establish the factors that will affect the final selection (the selection criteria), and
- evaluate the alternatives, in light of the established criteria.

Figure 8–5 (see page 142) shows that a comparative analysis can be written either objectively or subjectively. A writer following the left side of the diagram only *tells* about the alternatives and does not offer any opinions until reaching the recommendations. A writer following the right side of the diagram allows his or her opinions to be apparent right from the beginning. The primary difference is that, for an objective analysis the selection criteria are introduced *after* describing the various alternatives, whereas for a subjective analysis the selection criteria are introduced right at the start. Dan wrote his report using the objective approach.

⑯ Having described the two systems in previous paragraphs, Dan now identifies the key criteria he will use to assess the suitability of each system. In his case these are

- the desired level of customer convenience,
- how long the system will be needed, and
- the maximum acceptable cost.

Video Cameras

We have investigated available video cameras and have identified the VideoEye model VE401 as the most suitable for Provo's catalog order stations. It is robust, can be mounted permanently either above the video monitor or on the wall behind it, and is housed in a virtually indestructible shell. It has a visual search feature, an automatic variable-focus lens with a 2 in. to 100 in. range, an integrated microphone and miniature loudspeaker, a USB connector that plugs directly into the computer, and a 12-month warranty. For more details see the descriptive leaflet at Appendix A.

⑫

List price of the VideoEye VE401 is $129.95. However, we have negotiated bulk purchase prices of $88.90 each for 500 or more units and $81.15 each for 1000 or more units. Other video cameras we considered had comparable features but their list prices were 10% to 15% higher, and the lowest bulk purchases we were able to negotiate were $98.50 for 500 units and $95.50 for 1000 units.

Cost

The cost to install the voice or video equipment in the order centers will be much more expensive for video cameras than for telephones:

⑬

	Telephone	Video Camera
1 unit per order center	$ 5,703.05	$51,473.10
2 units per order center	$10,572.75	$92,754.45

However, the annual cost to rent dedicated telephone lines and to maintain the equipment will be much more expensive for telephones than for video cameras, because computer communication links already exist between the order stations and the customer service center:

	Telephone	Video Camera
1 unit per order center	$61,816.00	$12,353.54
2 units per order center	$86,205.00	$22,261.20

The total costs for the first year and each subsequent year of operation are summarized in Table 1. These costs are based on the following assumptions:

- The telephones or video cameras will be installed in 564 order stations and at 15 customer service positions.

- There will be a 2% per month failure rate, requiring equipment replacement.

Figure 8–5   The two plans for presenting a comparative analysis.

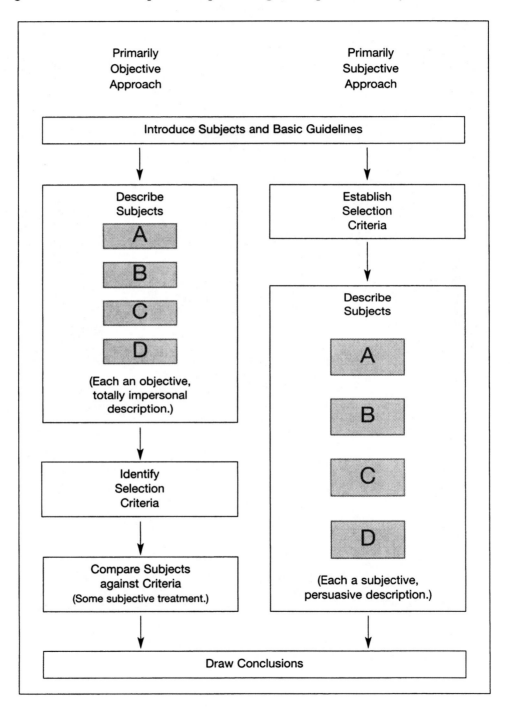

⑭

Table 1. Comparison of Equipment Purchase and Operating Costs		
Units Per Order Station	1st Year of Operation	Each Subsequent Year
1 Telephone	$67,519	$61,816
2 Telephones	$96,778	$86,205
2 Video Cameras	$115,015	$22,261
1 Video Camera	$63,827	$12,353

Detailed cost calculations are shown in Appendix B.

Evaluation Plan

⑮ There are three factors to consider when identifying the most suitable method to encourage customers to resume placing orders at Provo's catalog order centers:

1. Convenience to the customer. The most convenient setup will be to have a telephone or video camera for each order station, i.e. two per catalog order center.

⑯ 2. Length of time the voice or video communication system will need to be in place. The communication system will be needed a maximum of five years, by which time Provo plans to phase out telephone ordering altogether. However, we estimate that it may be possible to phase out the in-booth telephones or video cameras up to 2½ years earlier, depending on how rapidly customers become accustomed to ordering online.

3. Installation and operating costs. Provo has budgeted a maximum of $75,000 for 2002, and $20,000 per year for the next four years.

Evaluation of Alternative Methods

⑰ • **Method A — 1 Telephone per Order Center**—will be within Provo's budget by $7481 for year 2001, but will exceed the budget by $41,816 per year for years 2003 through 2006 and will not provide optimum convenience.

• **Method B — 2 Telephones per Order Center**—will provide optimum convenience but will exceed the budget by $21,778 for year 2002, and by $66,205 per year for each subsequent budget year.

• **Method C — 2 Video Cameras per Order Center**—will provide optimum convenience but will exceed the budget by $40,015 for year 2002, and by $2261 for each subsequent budget year.

• **Method D — 1 Video Camera per Order Center**—will be within Provo's budget by $11,173 for year 2002, and by $7647 for each subsequent budget year. It will not, however, provide optimum convenience.

5

(17) At point (10) on page 3 of the report, Dan presented the four alternatives as two groups: either one or two telephones, or one or two video cameras. Now, he reverses the sequence within the video option, so that the configuration he plans to recommend will be discussed last. (When presenting a comparative analysis, you have the choice of describing the alternatives in two ways: either start with the best choice and end with the least likely choice, or start with the least likely choice and end with the best choice. However, you do not tell your reader you are doing this; your reader should sense it subliminally.) Dan chose to work from the least-preferred choice to the best choice, and labeled the alternatives from A to D, to simplify referring to them later in his analysis.

(18) By placing a detailed description of the advantages and disadvantages of each alternative in a table, Dan provides an easy-to-refer-to overview. But he chooses to place it in an appendix to avoid interfering with reading continuity. He wants his readers to follow his analysis rather than become involved in studying the various factors and perhaps risk losing the thread of his argument. Consequently, he prefers that they refer to the table later, after they have read the report.

(19) Now Dan evaluates the alternatives, carefully describing how each does or does not meet the key criteria.

(20) Because none of the systems meets all the criteria, Dan has to decide whether his report should recommend exceeding Provo's budget or reducing customer convenience. He chooses the latter, because exceeding the budget would be extraordinarily expensive and would create considerable reader resistance. This means he has to persuade his readers that reducing customer convenience is not as significant as might be expected.

(21) Report writers who have done a thorough project or study will have a clearly defined project outcome in mind. If they are to convince their readers that the outcome they describe is valid and the action they suggest is viable, eventually they must cease being purely factual and start persuading their readers to agree with their results. At this stage they can let their subjectivity show and their opinions become apparent.

⑱ The advantages and disadvantages of each method are summarized in Appendix C.

⑲ Only methods A and D fall within the first-year budget criterion (by at least $7400), but only method D falls within the subsequent-year budget criterion (by at least $7600 per year). However, neither method meets the optimum-convenience criterion. To achieve optimum convenience for customers will mean placing two telephones or two video cameras in each catalog order center, one beside each order station, so that two customers can ask for assistance simultaneously.

⑳ We believe that occasions when two customers will require assistance at the same time will be minimal. The Dolman Survey Group's report shows that 30% of customers using the booths are already accustomed to ordering online and so will be unlikely to ask for help.[5] As time progresses, we expect at least 50% of customers who ask for help on their first visit will not need help on subsequent visits,[6] and that the need for a communication link will decline steadily. Consequently we suggest that the criterion requiring optimum convenience be relaxed to "reasonable convenience," by which we mean that every booth will have just one communication link, shared by the two order stations.

Should this link be a telephone handset or a video camera? Two factors need to be considered:

1. From the customers' viewpoint, a telephone is the medium with which they are familiar. Yet a video camera not only will demonstrate to them that Provo is moving to new technology but also will provide both visual and audio contact between the customer and the service representative.

㉑ 2. From Provo's viewpoint, the telephone handsets will be much cheaper to purchase and install than video cameras (see Appendix B). The cost of renting dedicated telephone lines, however, will bring the first-year telephone cost almost to that of the video camera purchase cost. As Figure 1 shows, in subsequent years the need to maintain dedicated telephone lines will cause the cost of using telephone handsets to far exceed the cost of using video cameras ($61,816 compared with $12,353 per year).

6

㉒ Illustrations should be chosen and inserted with care. They should

- serve a useful purpose,

- supplement, not duplicate, the written words,

- be simple, clear, and readily understood,

- be referred to in the narrative of the report,

- be accompanied by a brief caption or title, and sometimes a few explanatory remarks, and

- ideally, be smaller than a full page, so that some text can appear either above or below them (full-page illustrations tend to interrupt reading continuity).

More detailed information, particularly for preparing charts, graphs, and tables, is provided in Chapter 12.

Conclusions

㉓ The Conclusions and Recommendations are sometimes referred to jointly as a "terminal summary," meaning that they provide a summing-up of the outcome of the discussion. It's unwise, however, to treat them both under a joint heading, because doing so can invite a report writer to write a weak recommendation.

The most important thing to remember about Conclusions and Recommendations is that they must never offer surprises; that is, *they must present no new information.* Everything they contain must have been discussed in previous sections of the report (i.e. in the Discussion).

The Conclusions should

- be as brief as possible, with their main points drawn from the Discussion,

- be presented in descending order of importance, i.e. primary conclusion first, followed by subsidiary conclusions,

- satisfy the requirements established in the Introduction,

- never advocate action, and

- be presented in point form (in numbered or bulleted subparagraphs) if there are many subsidiary conclusions.

㉒

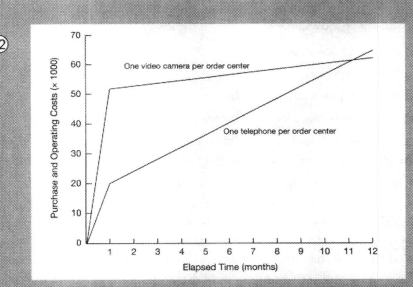

Figure 1. Comparison of Costs: Telephone vs Video Camera

㉓ **Conclusions**

If Provo is to rebuild confidence among customers who place orders from its mall-based online catalog order stations, it needs to install a communication link between the order stations and the company's customer service representatives. The communication link may be either voice-driven or video-driven. Customers are accustomed to telephone contact, but video cameras provide both face-to-face and audio communication.

Telephone handsets are much cheaper to purchase than video cameras, but their operation through dedicated telephone lines is much more expensive. The point at which purchasing and operating video cameras becomes less costly occurs between the eleventh and twelfth months of operation.

Ideally, there should be one telephone or video camera for each order station; that is, two per order center. However, the cost of installing and operating two telephone handsets per order center will exceed the budget by over 29% the first year, and by 331% in each subsequent year. For video cameras, the cost will exceed the budget by 40% the first year and by 11% each year thereafter.

For only one telephone set per order center, the cost will be 10% under-budget the first year and 209% over-budget in subsequent years, whereas for one video camera per order center the cost will be 15% under-budget the first year and 38% under-budget in subsequent years.

The convenience to customers may be less in a one-telephone or one-video camera catalog order center, but the inconvenience will diminish over the first 6 months as customers become accustomed to ordering online and their need for help is significantly reduced.

Recommendations

(24) The Recommendations should

- be strong and advocate action,

- use the active voice,

- satisfy the requirements established in the Introduction,

- follow naturally from the Conclusions,

- offer recommendations either in descending order of importance or in chronological sequence if one recommendation naturally follows another, and

- be in point form if several recommendations are being made.

To make a strong recommendation, you should write "I recommend ..." (if you are personally making the recommendation) or "We recommend ..." (if you are making recommendations for a group of people, a department, or your company). Never use the weak, passive voice for a recommendation, as in "It is recommended that ... ."

References/Bibliography

(25) A list of references or a bibliography lists all the documents the report writer used while conducting the project. Each reference describes the source of a particular piece of information, in sufficient detail so that readers can identify and obtain the document if they want to refer to it.

There are two main differences between a list of references (which Dan uses) and a bibliography:

- References are numbered and appear in the sequence in which each piece of information is referred to in the report.

- Bibliography entries are not numbered and appear in alphabetical sequence of primary authors' names. They may not be directly referenced within the report, but can be inserted to present the reader with a greater number of resources.

Generally, a list of references is more common in business and technical reports, and bibliographies are seen more often in professional journals and academic theses.

Chapter 11 contains suggestions for writing a list of references or a bibliography.

㉔   **Recommendations**

We recommend that Provo Department Stores installs one VideoEye model VE401 video camera in each of its 564 online catalog order centers, and a video camera above each of its 15 customer service video terminals. The total cost will be $51,473. We also recommend that Provo budgets $12,353 per year for maintenance and/or replacement of damaged or defective video cameras.

㉕   **References**

[1] *Study of User Reaction to, and Dissatisfaction with, Provo Department Stores's Computerized Independent Order Stations*, Report No. 0016, Dolman Survey Group, Buffalo, New York, January 14, 2001.

[2] Dolman Survey Group, p. 10.

[3] Dennis Hapgood, *Meeting Customer Needs in the New Millennium* (New York: Whitstable Press, 1999), p. 116.

[4] Hapgood, p. 174.

[5] Dolman Survey Group, p. 5.

[6] Hapgood, p. 177.

8

Appendix

(26) The Appendix contains complex analyses, statistics, manufacturers' data, large drawings and illustrations, photographs, detailed test results, cost comparisons, and specifications, i.e. any information that, if included in the discussion, would interrupt reading continuity. Often an appendix will contain detailed evidence to support what is said more briefly in the discussion. And sometimes the appendix section contains more pages than all the remaining sections of the report put together.

Certain guidelines apply to appendixes:

- Arrange them in the sequence in which each is referred to in the report.

- Give each appendix an informative title.

- Paginate each appendix separately, starting at "1" (except for single-page appendixes, which do not bear a page number).

- Assign an identifying letter to each appendix, starting with Appendix A, Appendix B, etc.

- Ensure that every appendix is referred to in the report narrative.

Because each appendix is a separate document, sometimes it may contain its own references. The references usually are listed at the end of the appendix.

Appendix A

Specifications:
VideoEye Model VE401 Video Camera

This was the manufacturer's 4-page
leaflet describing the miniature video
camera. It has been omitted from
this report to save space.

Appendix B

Cost Calculations

Costs for a Voice Communication System

	Cost per Unit	Cost for 564 Order Centers	Cost at Customer Support	Total Cost
1 Telephone per Order Center				
Telephones – Initial purchase cost	$9.85	$5,555.40	$147.75	$ 5,703.05
Maintenance/replacement (2%/mo)		$ 111.10	$ 2.95	$ 114.05/mo
Dedicated telephone lines (rent/mo)	$8.70	$4,906.80	$130.50	$ 5,037.30/mo
Monthly replacement/rental cost				$ 5,151.35/mo
Annual replacement/rental cost				$61,816.00/yr
Total Cost: Year 1				$67,519.05
2 Telephones per Order Center				
Telephones – Initial purchase cost	$9.25	$10,434.00	$138.75	$10,572.75
Maintenance/replacement (2%/mo)		$ 208.68	$ 2.77	$ 211.45/mo
Dedicated telephone lines (rent/mo)	$6.10	$ 6,880.80	$ 91.50	$ 6,972.30/mo
Monthly replacement/rental cost				$ 7,183.75/mo
Annual replacement/rental cost				$ 86,205.00/yr
Total Cost: Year 1				$ 96,777.75

Costs for a Video Communication System

	Cost per Unit	Cost for 564 Order Centers	Cost at Customer Support	Total Cost
1 Video Camera per Order Center				
Video Cameras – Initial purchase cost	$88.90	$50,139.60	$1,333.50	$ 51,473.10
Maintenance/replacement (2%/mo)				$ 1,029.46/mo
Annual replacement cost				$ 12,353.54/yr
Total Cost: Year 1				$ 63,826.64
2 Video Cameras per Order Center				
Video Cameras – Initial purchase cost	$81.15	$91,537.20	$1,217.25	$ 92,754.45
Maintenance/replacement (2%/mo)				$ 1,855.10/mo
Annual replacement cost				$ 22,261.20/yr
Total Cost: Year 1				$115,015.65

Appendix C

Comparison of Four Communication Alternatives

Factor/Feature	Method A: 1 Telephone	Method B: 2 Telephones	Method C: 2 Video Cameras	Method D: 1 Video Camera
Simplicity/familiarity for users	No learning needed	No learning needed	Some learning necessary	Some learning necessary
Contributes to Provo's "modern image"	No	No	Yes	Yes
Convenience/availability	Moderately convenient; available to only one customer at a time	Very convenient; continually available to customers	Very convenient; continually available to customers	Moderately convenient; available to only one customer at a time
Additional connection lines needed	1 dedicated telephone line	2 dedicated telephone lines	None needed	None needed
Extra support staff needed	Yes	Yes	Yes	Yes
Additional training for support staff	Internet ordering technology	Internet ordering technology	Internet ordering technology, plus screen management of video image	Internet ordering technology, plus screen management of video image
Cost: first year (2002)	$67,519	$96,778	$115,015	$63,827
Cost per year: subsequent years (2003–2006)	$61,816	$86,205	$ 22,261	$ 12,353

Dan Rogerson's Report Writing Sequence

The sequence in which Dan wrote his report is worth examining, because it shows that a long report does not have to be written sequentially.

1. First, Dan assembled the manufacturers' catalogs and specifications (he did not assign them appendix labels, because he did not yet know which he would use and where they would fit within the report).

2. Then he prepared the technical details that would support his discussion and evaluation, comprising

    *   the table of measurements, which later became Appendix B,

    *   the graph, which became Figure 1, and

    *   the comparison chart depicting the advantages and disadvantages of the four alternative systems, which became Appendix C.

3. He wrote the two sections describing the setup of equipment, and the section titled "Equipment Selection and Cost."

4. He wrote the Evaluation Plan, because it would provide the criteria against which he would compare each proposed system, and the comparative analysis ("Evaluation of Alternative Methods").

5. To set the scene, he wrote the Introduction and the section titled "Increasing Customer Acceptance of Online Ordering."

6. He wrote the Conclusions and Recommendations, which he found could be stated briefly because they were well supported by the previous sections (they simply summed up the key points in the Discussion).

7. Finally, he wrote the Summary.

8. And then, when the report was complete, he wrote the cover letter/executive summary.

PART 5

Report Writing Techniques and Methods

CHAPTER 9

Appearance and Format of Memorandum, Letter, and Semiformal Reports

The appearance of the reports you write should demonstrate the quality of your words. A neatly presented report, in a format that suits the circumstances and the intended reader, will create the impression that you are presenting valuable information. In readers' eyes it will enhance your credibility as a reporter of information, even before they start reading.

Conversely, a messy-looking report will create the impression that you are a careless writer whose information is of doubtful accuracy. And readers will gain this impression as soon as they pick up your report, before they read a word.

This chapter will help you choose the proper shape for each report you write and present it in a way that will subtly encourage readers to accept your facts and figures. There are four formats to choose from:

- A **memorandum**, which is used when a report is directed from one person to another within the same organization. The memorandum is the most informal form of report presentation.

- A **letter**, which normally is used when the writer of the report belongs to one organization and the person to whom it is directed belongs to another. Letters are more formal than memorandums but are still an informal reporting medium.

- A **titled document**, in which the report's title and the author's name are centered at the head of the first page, with the report narrative starting beneath them. Because its appearance is slightly more formal than that of a letter report, a titled document is often referred to as a *semiformal report*.

Figure 9–1   The shape of an informal memorandum report.

H. L. Winman and Associates

Interoffice Memorandum

To: W. K. Carter

From: R. Bryant

Date: January 27, 2002

Subject: Format for memorandum reports

Informality should be the keyword for your memorandum reports, both in appearance and language. The appearance should be simple, with the salutation and signature blocks omitted.

Informality, however, should not be interpreted as a signal that you can use sloppy language. Your reports must be organized coherently; your paragraphs and sentences must be properly constructed; and the words you use must be simple and clear.

If a report is long you may insert headings to help readers see how you have organized your information. You may also use subparagraphs or a bulleted list to break up long paragraphs or to list a series of points, as shown below:

- The subject line should be *informative*, so that it gives a clear indication of the memorandum's contents. For example: "Memorandum reports" would not have been sufficiently informative as a subject line for this memorandum.

- The first line of each paragraph may be indented 0.4 in. (12 mm) or started flush with the left margin. No indentation is more common.

- The names of persons who are to receive courtesy copies of the memorandum may appear directly under the addressee's name or entered three lines below the last paragraph, against the left margin (as has been done here).

You may sign or initial a memorandum report below the last printed line, as has been done here, or you may simply write your initials beside your printed name on the "From:" line at the top of the page.

Rod Bryant

c: L. L. Sampson

- A **bound document**, with a cover and full title page preceding the report proper, and separate pages for individual sections such as the Summary and Table of Contents. Such reports are known *as formal reports.*

Sample Reports

Guidelines for presenting informal and semiformal reports in the correct format are outlined in the sample memorandum, letters, and the first page of a semiformal report illustrated in Figures 9–1 through 9–5 of this chapter:

Figure	Page	Type of Report
9–1	157	Memorandum
9–2	159	Full block letter
9–3	160	Modified block letter
9–4	161	Traditional semiformal report
9–5	163	Contemporary semiformal report

Guidelines for presenting a formal report are included with the report analysis in Chapter 8.

Notes about Figures 9–2 and 9–3

(1) It is customary to name the person to whom a letter is addressed first and to follow the person's name with his or her title and then the name of the company or organization. The position the person holds may be placed beside the person's name (as in Figure 9–2) or on the second line (Figure 9–3).

(2) The *full* name and address of the recipient should appear in the letter because, with word processing, the address block often is highlighted and used to print the envelope.

Punctuation is omitted from the recipient's address, except where a comma is needed to separate two unrelated words in the same line. Although it is more common to insert a period after Mr., Ms., and a person's initials, there is a trend to omit such punctuation (it is omitted from the full block letter in Figure 9–2 but retained in the modified block letter in Figure 9–3). With electronic transmission of information, commas and periods may be either misread or converted into another symbol.

The correct way to type the city, state, and zip or postal code is to place them all on one line with the following spacing: City (1 space) two-letter State (2 spaces) zip or postal code. The US Post Office requirements state there should be no punctuation on this line.

Figure 9–2   The shape of a letter report in full block format.

OnLine Communicators Inc.
2820 Thurlow Street
Batavia NY 14022
Tel: 716-488-7125
Fax: 716-488-7294
olc@frontier.net

January 12, 2002

(1) R. Craig Williams, Head
Corporate Resources Division
Centaur Corporation
(2) PO Box 2760
Baton Rouge, LA 71036

(3) Dear Craig:

(4) **The Full Block Letter Format**

My analysis of 300 major companies in the US and Canada shows that 276, or
92%, prefer the full block letter style for their corporate correspondence and informal letter reports.

In the full block format every line starts at the left margin, which simplifies typing.
However, shorter (one-page) letter reports have to be carefully centered vertically
on the page if they are to achieve a balanced appearance.

Because it conveys the impression of a modern, forward-thinking organization, I
recommend you adopt the full block style for the Centaur Corporation's correspondence.

(3) Regards,

Marilyn P Duvall

Marilyn P. Duvall, Specialist
Business Communication

(5) enc

Figure 9–3   The shape of a letter report in modified block format.

Vancourt Business Systems Inc.

November 27, 2001

① Mavis J. Morgan
Manager, Customer Services
Cameron Manufacturing Inc.
② 2820 Border Road
Seattle WA 98016

③ Dear Ms. Morgan:

④   **Letter Reports in Modified Block Format**

   The modified block format is an older, more conservative and less used letter
style. When using this format

- type the name and address of the person you are writing flush with the left margin ("flush" means all lines start at the margin),

- start the first line of each paragraph either at the margin or indented 0.4 in. (12 mm), as has been done here,

- set the date to end flush with the right margin, and start the signature block at the page centerline,

- center the subject line and print it in bold characters, and

- ensure the left and right margins are roughly the same width.

   If the report is short enough to fit on one page, position it vertically so that the body of the letter is roughly in the middle of the page.

                               Sincerely,

③   *Martin Cartwright*

                           Martin Cartwright
                           Publications Editor

⑤ enc

③ The use of a colon (:) after the salutation and a comma (,) after the complimentary close is optional, but their insertion or deletion should be consistent.

④ Subject lines are optional. They should be in boldface type, and upper- and lowercase letters.

⑤ "enc" means that enclosures or attachments are being sent with the letter. Sometimes the enclosures/attachments are listed below or immediately after "enc," or the number of enclosures is listed: enc 3.

Notes about Figure 9–4

① The title should be set in boldface upper- and lowercase type and centered.

② The author's name and affiliation may appear either here or on the last page of the narrative (i.e. ahead of the attachments).

③ The first line of each paragraph may be indented 0.4 in. or may start at the left margin, as in Figure 9–4. The latter method is preferred.

④ To create an uncrowded appearance there should be one and a half or two blank lines between paragraphs but only one blank line between a heading and the paragraph that follows it.

Figure 9–4   A semiformal report set in one column (top of first page only).

① **Recommended Format for Semiformal Reports**

② Rodney T. Elson
Technical Communication Consultant

Summary

③ The appearance of a semiformal report lies halfway between the comfortable informality of the letter report and the strict formality of the formal report. The report's title is given prominence by being displayed in capital letters across the upper center of the first page, and the author's name and company affiliation are centered beneath it. The narrative of the report follows immediately and continues onto subsequent pages. The impression gained by the reader on first seeing the report should be of a high-quality document containing important information.

The Report's Parts

④ The parts of a semiformal report are similar to those of a formal report, except that the report normally has no cover, the summary seldom has a page to itself, and the table of contents page is omitted. As in a formal report, each major section is introduced by a center or side heading.

Improving the Body of the Report

You can help your readers more readily understand and access information by introducing *Information Design* techniques into your reports and proposals. Good *Information Design* helps readers focus their attention on specific information. The techniques include using imaginative design of the page, making careful use of headings, choosing an appropriate font, and choosing not to justify the text at the right-hand margin.

Redesigning the Page

In Figure 9–5 the semiformal report in Figure 9–4 has been redesigned so that the information is carried in two columns. The column on the left is about 1.75 in. (45 mm) wide and carries only the headings. The column on the right is 4.25 in. (110 mm) wide and carries all the text, tables, and illustrations. There is also generous use of white space between the paragraphs. This arrangement

- provides a shorter scanning line and so is easier on the eye,
- simplifies searching for specific information, because the headings are easy to see, and
- offers space for the reader to make notes beside the text.

We recommend this design for semiformal reports and proposals that are not written as letters or memorandums. For an example, see Leo Cheng's proposal on pages 91 to 104 of Chapter 7.

Choosing a Font

A font is a particular set of printing type with each letter having the same features; for example, **Century Schoolbook**, **Arial**, and Times New Roman. Fonts fall into two main types, known as *serif* and *sans-serif*:

- A *serif* font has little finishing strokes on the extremities of each letter; for example: T, E, L. The most well-known serif font is Times New Roman. The text in this book is set in a serif font known as Sabon.
- A *sans-serif* font is plain and has no finishing strokes; for example: **T, E, L**. Typical sans-serif fonts are **Arial** and **Helvetica**.

Generally, a serif font is preferred for long documents, whereas a sans-serif font is suitable for short documents and notices. Many people prefer the sans-serif font because it is clean and uncluttered, yet tests have shown that a serif font is easier on the eye

Figure 9–5   An alternative semiformal report

Alternative Format for Semiformal Reports

Rodney T. Elson
Technical Communication Consultant

Summary   The appearance of a semiformal report lies halfway between the comfortable informality of the letter report and the strict formality of the formal report. The report's title is given prominence by being displayed in large, boldface, upper- and lowercase letters across the upper center of the first page, and the author's name and company affiliation are centered beneath it.

The narrative of the report follows immediately and continues onto subsequent pages. The impression gained by the reader on first seeing the report should be of a high-quality document containing important information.

The Report's Parts   The parts of a semiformal report are similar to those for a formal report, except that the report normally has no cover, the summary seldom has a page to itself, and the table of contents page is omitted. As in a formal report, each major section is introduced by a side heading, and

because the finishing strokes help lead the eye from one letter to the next. For that reason, most books are set in a serif font. If you know your information will be read only online, we suggest using a sans-serif font, which can ease eyestrain that may be caused by the computer monitor.

When you choose a font, use it throughout the document. To show emphasis, use **bold**, *italic*, or larger characters (this is particularly useful for showing the different levels of headings). The only exception is the title of the document itself, which often is set in a bold sans-serif type to give it prominence. Avoid underlining because the underline appears too close to the characters and makes reading difficult.

What font size should you choose for the text? For serif fonts we recommend 12-pt type. For sans-serif fonts we recommend 11-pt type, or even 10 pt. (The sans-serif fonts look larger than serif fonts when set in the same point size.)

Justifying Text Only on the Left

Word-processing provides us with an easy way to make every line the same length, and so "justify" the right margin (i.e. make it straight). Visually, right-margin justification seems appealing. Yet, because it spreads the words and letters in some lines farther apart, and scrunches the words in other lines close together, justification can create an uneven, disturbing effect. Consequently, to improve readability, we recommend you create a ragged-right margin for your reports and proposals.

In books, however, which originally were typeset with "hot type," the spacing between letters could be controlled much more readily and right-margin justification became the normal way to present text. Therefore, we have asked the editor of this book to set this paragraph and the one above ragged right, so you can see the difference.

Avoiding All Caps

We recommend you avoid using all capital letters because they make your text more difficult to read. This applies equally to headings: instead, use boldface upper- and lowercase. Compare the following:

**COMPARISON OF WATER LEVELS IN LAKES AND RIVERS OF
WISCONSIN, MINNESOTA, AND NORTH AND SOUTH DAKOTA**

**Comparison of Water Levels in Lakes and Rivers of
Wisconsin, Minnesota, and North and South Dakota**

The solid block of letters in the upper example makes dull, unemphatic reading. The "ups and downs" of the letters in the bottom example catch reader interest and the eye flows easily from word to word.

Using Tables to Display Information

A table does not only have to carry numbers. Sometimes you can present information better in a table rather than as straight text. Here are two examples of the same information:

Travel Expense Guidelines No. 1

Certain guidelines apply to travel expenses. All air travel must be booked with Haynes Travel Services and the cost charged to account A78641. Attach Haynes's invoice and the ticket stub to your expense claim. Request hotel/motel accommodation at our corporate rate (quote file 2120), pay with a company Visa card, and attach hotel receipt. The per diem rate for meals is $30. No receipts are necessary except for meals over $25 (excluding tip).

Travel Expense Guidelines No. 2

Air Travel	• Book with Haynes Travel Services • Charge to Account A78641	• Haynes's invoice • Ticket stub
Accommodation	• Request corporate rate (quote file 2120) • Pay with company Visa card	• Hotel/motel receipt
Meals	• Per diem rate is $30	• Receipt not required unless meal cost is over $25 (excluding tip)

The information presented in the table is far easier to access than that in the paragraph.

CHAPTER 10

Developing a Writing Style

We use essentially the same language for report writing as we do for regular correspondence, and in both cases we strive to be brief yet fully informative. A report must contain all the information its readers need to understand a given situation and, if necessary, take action. Yet it must neither waste the readers' time by conveying too many details nor obscure the message by using ponderous sentences and paragraphs. The tone, style, and language of the report must be your own. Your readers should be able to hear your voice and sense your personality behind your writing.

This chapter describes techniques that will help you focus your reports correctly, be direct, and avoid cluttering the narrative with unnecessary words and expressions.

Get the Focus Right

The first rule of report writing is never to start writing until you have answered three questions:

I.   Who is my reader?

2.   What is the purpose of my report?

3.   Do I want to be purely informative or convincingly persuasive?

Your answers will give you a sense of direction and help you write much more easily and spontaneously than if you had simply picked up a pen or placed your fingers on the keyboard and started writing. The implications of each question are discussed on the following pages.

Identify the Reader

Kim Wong has been studying materials-handling methods used by her company and feels there is a better way to manage the ordering, receiving, documenting, storing, and issuing of parts and materials. Her department manager, Anna Sharif, has told Kim that she is to write a report of her findings and recommendations but has not told her who will be reading the report and using the information it contains.

Kim is likely to make many false starts if she tries to write without a clearly defined audience in mind. Without a known target, she will write in a vacuum and will be unable to focus her report. She cannot assume she is writing only for Anna's eyes, since Anna may be planning to send the report directly to the head office with an accompanying memorandum. Kim's report would then be read by a much wider audience. To focus her report properly, Kim needs to find out not only who is going to read it but also how it will be used.

To help you identify the reader, review Chapter 2 to determine the specific questions you need to answer.

Identify the Purpose

Always check that you have clearly identified the purpose of your report before you start writing. On a separate sheet, write

"The purpose of my report is to ..."

and then complete the sentence (but limit yourself to only one sentence). Kim Wong, for example, should write

The purpose of my report is to demonstrate that our materials-handling methods are outdated and to show how they can be improved.

Kim's next step is to decide whether her report is to be informative or persuasive (i.e. if it is to tell or to sell). When a report writer is simply informing readers of a given situation, then the sole purpose of the report is to present facts; but if the writer wants to evoke a response to the situation, then the report must convince readers that the writer has a valid point and persuade them to act. If Kim knows that the department manager will be her only reader and that Anna is seeking only facts, then she will write an informative report telling her what she has found out. Alternatively, if she has discovered that her report will be sent to the head office to convince executives to invest in a new materials-handling system, she will have to write a persuasive report because she has to "sell" the new system to the head office staff; she expects a response or action from the reader.

Write to Inform

Informative writing is much simpler than persuasive writing. To write informatively you need to present facts clearly and in a logical sequence. You should write briefly and directly, and closely follow the pyramid structure described in Chapters 2, 3, and 4. The interrupted data transmission tests described in the report on page 21 and the mobile trailer progress report on pages 46 and 48 are typical examples of informative writing.

After reading an informative report, the reader's response should be simply "Okay, thanks for letting me know."

Write to Persuade

The difficulty with persuasive writing is preserving one's objectivity. Although your aim should be to convince readers to accept your ideas and take action, your partiality or bias should not be so obvious that readers feel they are being coerced. The big picture is to persuade, but you cannot be convincing unless you provide solid facts that inform.

Fortunately, several sections of every persuasive report deal with facts, and these can still be presented informatively. These are the Introduction, in which you describe the background to your report, and the description of your Approach and Findings. (Kim Wong's description of the existing materials-handling methods, for example, should be strictly informative, regardless of whether her report is informative or persuasive.) Only when you have to present your suggestions and analyse advantages and disadvantages should your involvement become obvious. Of course, when you make a recommendation your preference should be readily apparent.

The suggestion and proposal in Chapter 7 and the formal report in Chapter 8 are examples of persuasive writing. The two longer reports, particularly, show how their authors have gradually developed their cases, working carefully from an informative presentation of facts toward a persuasive evaluation of alternatives.

Be Direct

Chapters 2 through 8 stress the need to satisfy a reader's curiosity by identifying the most important information, consolidating it into a short summary statement, and placing it at the front of every report you write. This "direct" writing technique can be extended to individual sections of a report and to each paragraph. It can also be enhanced by writing as much as possible in the first person and in the active voice.

Use the Pyramid Structure

There is no need to feel that the pyramid structure described in Chapter 2 applies only to a complete document. Within a long report each major section should also be structured "pyramid style," with each section opening with a short summary statement followed by the basic BACKGROUND-FACTS-OUTCOME arrangement of information. This technique is even used in this chapter: the opening paragraph of each major subsection starts with a summary statement. For example, the heading "Get the Focus Right" on page 166 is followed immediately by this short paragraph:

> The first role of report writing is never to start writing until you have answered three questions:
>
> 1. Who is my reader?
>
> 2. What is the purpose of my report?
>
> 3. Do I want to be purely informative or convincingly persuasive?

These five lines identify the topics discussed in the remainder of the subsection.

Similarly, the first paragraph of *this* subsection (immediately following the heading BE DIRECT) summarizes what is being described here and on the next few pages.

The pyramid can even be used to structure paragraphs. The first sentence becomes the topic sentence (Summary Statement), and the remaining sentences support and develop the initial statement. For example, in the two paragraphs that follow, the topic sentences (in italics) describe the main point while the remaining sentences provide more details.

> 1. *We have evaluated the condition of the Merrywell Building and find it to be structurally sound.* The underpinning done in 1958 by the previous owner was completely successful and there still are no cracks or signs of further settling. Some additional shoring will be required at the head of the elevator shaft immediately above the 9th floor, but this will be routine work that the elevator manufacturer would expect to do in an old building.
>
> 2. *The costs of poor communication are seldom calculated but should never be overlooked or simply brushed aside.* An inadequately worded purchase order that results in the wrong goods being delivered can increase the cost of doing business and involve countless people in correcting the error. Inadequate directions for an appointment, which cause one or more persons to go to the wrong place and to waste time, are equally costly. So are incorrect facts or ambiguous calculations that create disagreement—if not outright conflict—between the transmitter and receiver of the information.

In high school and college or university you were probably told that every paragraph must have a topic sentence. For report writing—and particularly for short reports—you should place your topic sentence at the beginning of each paragraph. This pyramidal approach helps your readers to discover quickly what each paragraph is about and to easily grasp the details that follow. Readers particularly appreciate this technique when they only have time to skim a document yet need to understand its content.

If you have to prove your case to readers who may be prejudiced against or tend to resist the facts you have to present, a nonpyramid paragraph may be more suitable. Even then, it should only be used occasionally to create a particular effect. For example, in the following paragraph, the events lead the reader up to the main point the author wants to make (it's in the italicized topic sentence).

> We first monitored sound levels between 8 p.m. and 10 p.m. to establish a background sound level while the building was empty. Then the following day we measured sound levels hourly from 7 a.m. to 6 p.m. at 28 locations throughout the building and recorded the results in Appendix B. Seventeen of the measuring points were in or adjacent to departments where employees had complained of excessively high noise, and 11 were control points in departments from which no complaints had been received. *Although the sound levels measured at the "noisy" locations were an average 7 decibels higher than at the control locations, at no location was the sound level higher than 63 decibels.*

In this example, the climactic method (leading up to the main point) is used to help persuade, although there is a risk the reader will get lost or bored with all the details. But the same information can be presented just as easily in the pyramid style to readers who are better prepared to accept the facts.

> *We monitored sound levels and found that they did not exceed 63 decibels anywhere in the building, although the sound level for departments reporting excessively high noise was an average 7 decibels higher than the sound level for other departments.* Our measurements were recorded on two separate occasions: between 8 p.m. and 10 p.m. one evening to establish a background sound level while the building was empty, and hourly from 7 a.m. to 6 p.m. the following day. Seventeen of the 28 measuring points were in or adjacent to departments where employees had registered complaints, and 11 were control points in departments from which no complaints had been received. The results are shown in Appendix B.

The approach you use will depend on the effect you want to create.

Write in the First Person

Reports are written and read by people, so it is natural to write from person to person. Yet many report writers try to avoid using the first person when they write because they feel they are being "unbusinesslike" or "unprofessional," or they have to remain totally objective. They write

The components have been ordered ...

A data survey was conducted ...

A report was submitted ...

It is recommended that ...

The writers of these statements seem afraid to say they were involved in ordering the components, the data survey, the report submission, and the recommendation. Their statements would be much more direct and effective if a "person" could be inserted into them.

I have ordered the components ...

We have conducted a data survey ...

I submitted my report ...

We recommend that ...

To write in the first person (i.e. to use *I, we, me,* and *my)* is not unprofessional or unbusinesslike. When you write a report to your manager, to someone in another department, or to someone outside your own organization, you should try to write from person to person. Insert *I* if you are writing for yourself, and *we* if you are reporting for a group of people, your department, or your company.

In today's technological world it is important we put people back into our communications. Too often we read reports about departments or offices performing some action: *The office will process the form and return them to the originator.* These are inanimate objects. People should be mentioned when they are the ones performing the work: *We will process the forms and return them to you.*

If you are drafting a report for another person's signature (your department head, for example), you may feel you do not have the right to use *I* and *we*. Under these circumstances you should go to the person whose name will appear on the report and ask if you can use the first person.

There are numerous examples of reports written in the first person throughout Chapters 3 through 8. The pronoun *I* is readily evident in the informal memo reports in Chapters 3 and 4 (in particular, see Frank Crane's trip report on page 25, Marjorie Franckel's progress report on page 43, and Tom Westholm's investigation report on page 54). Only in the slightly more formal inspection report on page 33 and progress report on pages 46 and 48 is *I* less evident: Paul Thorvaldson has limited its use primarily to his suggestions (his Outcome compartment), and Roger Korolick uses *I* only when he refers to his concerns and plans.

In the longer reports, the first person plural (*we*) is used by both Leo Cheng in his in-house proposal on pages 91 through 104, and Tod Phillips in his client-oriented investigation report (pages 73 through 79). Dan Rogerson predominantly uses *we* in his formal evaluation report on pages 129 through 153; he limits *I* to moments when he is clearly making a personal observation.

Use the Active Voice

If you were presenting this information, which way would you write it?

A. Carl Dunstan investigated the problem.

B. The problem was investigated by Carl Dunstan.

In report writing you should try to be as direct and as brief as possible without losing any information. Since both sentences contain the same information, your choice should be sentence A because it is shorter and more direct.

Sentence A is written in the active voice, in which the person or object performing the action is stated first:

Carl investigated the problem.

The shaft penetrated the casing.

Petra is studying the charts.

Sentence B is written in the passive voice, in which the person or object performing the action is stated *after* the verb:

The problem was investigated *by Carl.*

The casing was penetrated *by the shaft.*

The charts are being studied *by Petra.*

Sentences written in the active voice are generally shorter and more emphatic than sentences written in the passive voice. Similarly, reports written primarily in the active voice seem much stronger, more definite, and more convincing than reports written predominantly in the passive voice. Your reader will also be able to comprehend and retain the information more easily.

Compare the following two paragraphs, both describing the same situation:

Primarily Passive Voice:

A study of electricity costs was conducted in three stages over a twelve-month period. First, a survey was taken and a list made of all apartment dwellers in the area. Then a table was constructed in which family size was compared against apartment size. Finally, an analysis was made of apartment dwellers' lifestyles and their major appliance ownership (it was assumed that a stove, refrigerator, and air conditioner were installed as standard equipment in each apartment). *(77 unassertive words)*

Primarily Active Voice:

We studied electricity costs in three stages over a twelve-month period. First, we surveyed and listed all apartment dwellers in the area, and then constructed a chart comparing family size against apartment size. Finally, we analysed apartment dwellers' lifestyles and their major appliance ownership (we assumed that each apartment was equiped with a stove, refrigerator, and air conditioner as standard equipment). *(62 confident words)*

The information conveyed by the two paragraphs is the same, yet the impact each creates is markedly different. The active voice paragraph appears to be written by a confident, knowledgeable individual who uses the first person (we) and clearly identifies that someone has been actively doing something. The passive voice paragraph appears to be written by someone who is detached and uninvolved; its author does not write in the first person and so writes sentences without mentioning who performed the action. This creates the impression that he or she is merely passing along information.

Writing in the first person can help you avoid writing in the passive voice. When you write *I* or *we,* you immediately identify who was involved:

I requested approval to visit ...

Early in May *we* established criteria ...

This becomes particularly important when you have to make recommendations. The Recommendations section of a report must be strong and definite; yet often report writers adopt an indefinite, passive stance, writing

It is recommended that ...

Instead, they should be firm and assertive, and write in the active voice:

I recommend ...   (when the report writer is making a recommendation as an individual), or

We recommend ...   (when he or she is making a recommendation on behalf of a group of people, the department, or the company).

Both Leo Cheng in his proposal, and Dan Rogerson in his formal report, write "We recommend ..." in the Recommendations sections (see pages 103 and 149).

Writing in the active voice does not mean you always have to write in the first person. You can just as easily name another person, a department, or an object.

Mr. Singh revised the estimate.

To meet the mailing deadline *the project group* worked until 3 a.m.

My bank increased the interest rate.

On the fourth floor *the video cart* lost a wheel.

Our comptroller recommended a budget cut.

After reading our report, *the client* requested a revision.

Of course, when you do not know who performed the action, prefer not to name names, or want to de-emphasize the doer, then the passive voice has to be used.

Your budget has been cut by 30%. *(Specific person not stated.)*

The documents were misfiled. *(Person not known.)*

The long-awaited Manston report has been printed. *(The emphasis would be wrong in the active voice: "The printer has printed the long-awaited Manston report. ")*

This section only draws your attention to the active voice and suggests you use it wherever possible in your reports. For a more detailed description refer to a language textbook, such as *The Prentice-Hall Handbook for Writers*.

Avoid "Clutter"

The words you use in a report can do much to help your readers understand quickly what you have to say, and then react or respond in the way you want them to. A clear, concise report will evoke the correct reader response, but a report cluttered with unnecessary words and expressions can so muffle the message that readers either miss the point or lose interest and stop reading. Because "clutter" words are used frequently by *other* people, we tend to recognize them as old friends and so may have difficulty weeding them out of our own writing.

Use Simple Words

When you have the choice between two or more words, try using the simpler word. The accountant who refers to "remuneration" and "superannuation scheme" in an annual report would do better to write about pay, salary or wages, and the pension plan. Then he or she would be understood by virtually every reader, from the chief executive to the newly employed warehouseperson. Similarly, the design engineer who writes that the modem he or she has developed "utilizes uniquely sophisticated circuitry" would sound less obscure if he or she said it "uses complex circuits."

There are certain words peculiar to each of our particular vocations that we have to use because no other words can adequately replace them (computer specialists, for example, refer to disks, disk drives, and bits and bytes of information). To keep their sentences and paragraphs as uncluttered as possible, specialists should surround these technical words with simple words.

Compare the following two sentences:

A. An aberration of considerable magnitude significantly influenced the character readout.

B. A large deviation seriously affected the character readout.

Readers would need a very good vocabulary to understand all the words in the first sentence, and even then they would have to read carefully to fully grasp what is being said. Most readers would understand the second sentence.

Remove Words of Low Information Content

If you want your writing to create a positive, purposeful impact, you cannot afford to insert words and expressions that neither clarify nor contribute to the message. Such words are known as *low information content* (LIC) words because they detract from rather than improve the message's clarity. There are several in the following sentence:

> In order to effect an improvement in package handling, an effort should be made to move the shipping department so that it is located in the vicinity of the loading dock.

Can you identify the LIC words?

> *In order to* (replace with *to*)
>
> *effect an improvement in* (use *improve*)
>
> *an effort should be made* (replace with *we should*)
>
> *located in the vicinity of* (use *nearer to*)

Without the LIC words the sentence reads

> To improve package handling we should move the shipping department so that it is nearer to the loading dock.

Or (better still),

> To improve package handling we should move the shipping department nearer to the loading dock.

Some common LIC words and expressions are listed in Table 10–1.

TABLE 10-1   Some Low Information Content (LIC) Words and Expressions

These LIC words and phrases should be eliminated (indicated by X) or written in a shorter form (shown in parentheses).

actually (X)
a majority of (most)
a number of (many, several)
as a means of (for, to)
as a result (so)
as necessary (X)
at present (X)
at the rate of (at)
at the same time as (while)
at this time (X)
bring to a conclusion (conclude)
by means of (by)
by the use of (by)
communicate with (talk to, email,
    telephone, write to)
connected together (connected)
contact (talk to, email, telephone,
    write to)
due to the fact that (because)
during the course of (during)
during the time that (while)
end result (result)
exhibit a tendency to (tend to)
for a period of (for)
for the purpose of (for, to)
for the reason that, for this reason
    (because)
in all probability (probably)
in an area where (where)
in an effort to (to)

in close proximity to (close to, near)
in color, in length, in number, in size (X)
in connection with (about)
in fact, in point of fact (X)
in order to (to)
in such a manner as to (to)
in terms of (in, for)
in the course of (during)
in the direction of (toward)
in the event that (if)
in the form of (as)
in the light of (X)
in the neighborhood of, in the vicinity of
    (about, approximately, near)
involves the use of (employs, uses)
involve the necessity of (demand, require)
is a person who (X)
is designed to be (is)
it can be seen that (thus, so)
it is considered desirable (I or we want
    to)
it will be necessary to (I, you, or we
    must)
of considerable magnitude (large)
on account of (because)
on the part of (X)
previous to, prior to (before)
subsequent to (after)
with the aid of (with)
with the result that (so, therefore)

LIC words make writing seem woolly and indefinite. They flow easily from our fingertips, and once they are on screen or paper they can be hard to identify. For example:

When we have written	*It can be difficult to think of*
brings to a conclusion	concludes
for a period of	during
it will be necessary to	we must
in the direction of	toward

Simply being aware that you should not use LIC words in your reports will help you to be a careful writer but still will not prevent you from inserting them inadvertently during an enthusiastic burst of writing. After you have typed your first draft but before the final copy is printed always take a few minutes to check that you have not used any unnecessary words.

Eliminate Overworked Expressions

Overworked expressions can create an even more noticeable negative effect than LIC words because they make the writing seem wordy or insincere and sometimes pompous or evasive. Some typical expressions are listed in Table 10–2. These should be searched for, identified, and eliminated at the same time as you check for LIC words.

TABLE 10–2   Overworked Expressions and Clichés

a matter of concern	in the long run
and/or	in the matter of
all things being equal	it stands to reason
as a last resort	last but not least
as a matter of fact	many and diverse
as per	needless to say
attached hereto	on the right track
at this point in time	par for the course
by no means	please feel free to
conspicuous by its absence	pursuant to your request
easier said than done	regarding the matter of
enclosed herewith	slowly but surely
for your information (as an introductory phrase)	this will acknowledge
	we are pleased to advise
if and when	we wish to state
in reference to	with reference to
in short supply	you are hereby advised
in the foreseeable future	

Avoiding Gender-specific Language

In today's diverse society we can no longer assume that the person who delivers our mail is a mail*man* or that the person leading the meeting is the chair*man*. As well, we cannot assume a secretary or nurse is female or that an engineer or lawyer is male. Our society has reached a point where more opportunities and career paths are open to all individuals regardless of their gender. Unfortunately, our language has not kept pace. Because we can unknowingly offend our readers, we need to alter our thinking and our writing style to avoid these misunderstandings.

Instead of writing

> The committee decided to contact an electrical contractor. *He* will visit the new office building and provide a cost estimate for installing an air-conditioning unit.

and

> The secretary will write the meeting minutes. *She* will also coordinate the travel arrangements and the meeting facilities.

Try writing

> The committee decided to have an electrical contractor provide an estimate for installing an air-conditioning unit in the new office building.

and

> The secretary will write the meeting minutes, coordinate the travel arrangements, and organize the meeting facilities.

Be Consistent When Referring to Men and Women

Usually men have used the courtesy title *Mr.* to precede their names. Until 20 years ago, women had two courtesy titles to denote whether they were married or single: *Mrs.* and *Miss.* (English is not the only language to do this; for example, in France, men are referred to as *Monsieur* and women as *Madame* or *Mademoiselle.*) Today, a woman's marital status is *never* implied in her title: all women should be referred to as *Ms.* Similarly, *never* address a letter to "Dear Sir or Madam."

Many job titles are just as gender-specific and predominantly male-oriented. These have been changed in recent years so that the title refers to both male and female employees. Table 10–3 lists gender-specific titles and suggests better alternatives.

TABLE 10-3   Preferred Terms for Gender-specific Titles

If you are tempted to write:	*Consider using:*
actor; actress	actor (for both sexes)
chairman	chairperson or chair
cowboy	cattle rancher
fireman	firefighter
foreman	supervisor
policeman; policewoman	police officer
postman	letter (or mail) carrier
repairman	service technician
salesman	sales representative
spokesman	spokesperson
workman	worker or employee
waiter; waitress	server (or waiter for both sexes)

Note: The term *man-hours* was once used to define the time that would be expended on a particular job. Today we write *work-hours* or *staff-hours*.

CHAPTER 11

Writing a List of References or a Bibliography

Whenever you quote someone else's facts and figures, or draw information from a text-book, journal article, report, letter, email, the Internet, the Web, or even a conversation, it is customary to acknowledge the source of your information within your report. This is usually done at the end of the report in a section called "References" (or "List of References"), as Dan Rogerson has done in his formal report in Chapter 8. References normally occur in longer reports and proposals—seldom in very short reports.

The purpose of a reference is threefold:

1. To give your report credibility. When readers encounter a statement such as "A recent study shows that 37% of color monitors emit radiation," they expect to be told who made the original statement and in what document it appeared.

2. To help readers refer to the same source if they want more information.

3. To give credit to the originator.

There are specific rules for writing a list of references, and to some extent they vary depending on where and in what form your report is to be published. The rules shown here are generally acceptable for any business or industrial technical report. We are assuming that most reports you write will be for your company or the organization that employs you, and that the standard style we present here will apply.

However, if you are writing a scientific paper or a report that will be published in the journal of a professional society, then you will need to adopt the style used by that particular journal. The journal may prefer that you write a bibliography rather than a list of references. Consequently, we also present some brief guidelines for writing a bibliography. For more information we suggest you refer to *The Chicago Manual of Style*,[1]

Kate Turabian's well-known *A Manual for Writers of Term Papers, Theses, and Dissertations*,[2] or the *MLA Handbook for Writers of Research Papers*[3] (MLA = Modern Language Association).

In examining these and other referencing guidelines you will notice that some tell you to <u>underline</u> the title of a document (this is done in the *MLA Handbook*, for example), whereas others tell you to set the title *in italics* (this is done in *The Chicago Manual of Style*). For simplicity, and especially in today's word-processing environment, we recommend you set document titles in italics, in both a list of references and in a bibliography. The reason? In the days when an author wrote a research paper by setting pen to paper or by using a typewriter, it was customary to underline the title of a document. This acted as a message to the publisher, in effect saying: "When you print these words, please set them in italics." Today, when almost everyone writes their reports and technical papers at a computer keyboard, the word-processing software permits you to set the words in italics *as you key them in*, and so saves a step in the publication process.

Where in previous years all source referencing was to printed documents or to a spoken observation, today it's becoming increasingly likely you will be referring to and listing an electronic source from which you gained information. An electronic source can be accessed in two ways. It may be on a CD-ROM, a 3.5-inch disk, or magnetic tape, in which case it is always available, much as a print document is continually accessible. Alternatively, a source may be online, such as on a Web page or the Internet, in which case its presence may be only transitory and there is no guarantee it will continue to be accessible. For that reason, online sources require more detailed referencing: you need to record not only the electronic source of the item, but also its original identification (i.e. as a book, journal article, newsletter, etc.). This will be demonstrated in the following pages.

How to Write References

References are listed in the order in which they appear in the report. If the first statement that needs to be supported concerns the quantity of water consumed by your city, then the first item in your list of references will be the document in which water consumption is tabulated. Each reference entry is numbered sequentially, starting at "1," and a corresponding number is shown in the report narrative to direct the reader's attention to the appropriate entry in the list of references. For example:

Over the past eight years the city's water consumption has ranged from a low of 207,389 gallons per day to a high of 253,461 gallons.[1]

(This superscript [1] refers to the first entry in the list of references.)

Since every entry is numbered, a corresponding number like this must appear in the report narrative for each reference. (See Don Rogerson's report, pages 137 and 139.)

Each entry in the list of references must supply certain primary information so that the reader can clearly identify the document and be able to refer to it or order it. For example, the report must identify

- who made the statement,

- in what document it appeared, or where and to whom it was said (if a spoken reference), and

- when the statement was made.

Every detail (such as the author's name and document title) must be copied *exactly* as it appears on the original document, so that readers will experience no difficulty in finding or ordering the document.

The preferred methods for listing the more common documents and oral presentations are described below. They are based on the guidelines provided by the Modern Language Association (MLA).

Book by One Author. The entry should contain

author's name,

book title (in italics),

city of publication,

name of publisher,   ⎤

                          ⎥   (enclosed within brackets)

date of publication,   ⎦

page number of specific reference (if applicable).

Here is an example:

1. Mavis Gerbrandt, *Neural and Fibreoptic Networks* (Los Angeles, CA: Technical Associates Inc., 2000), p 231.

(Note: The edition number is omitted for the first edition of a book.)

Book by Two Authors. Both authors are named; all other information is the same as for a single-author book. This book and page are an example:

2. Ron Blicq and Lisa Moretto, *Writing Reports to Get Results,* 3rd ed (New York: IEEE Press, 2001), p 182.

Book by Three or More Authors. Only the primary author is named (usually the first-named author); remaining authors are replaced by the expression "and others" (the expression *et al* is no longer used). All other information is the same as for a single-author book.

3. Peter L. Gneiss and others, *Introducing New Technology to the Developing Nations* (New York: Scientific and Technical Press, 1999), p 208.

An Anthology. An anthology is a book containing sections written by different authors, with the whole book edited by another person. If your reference is to the whole book, the editor's name is used and his or her editorial role is identified by the word "ed" immediately after the name.

> 4.   Christine L. Summakindt, ed, *Marketing in Pacific Rim Countries* (Portland, OR: Dover Books Inc., 2001), p 3.

If your reference is only to an article or section in the book, the author's name and section title are used, so that the entry contains

author's name (or authors' names),

section title (in quotation marks),

book title (in italics),

editor's name (if the book has an editor),

city of publication,

name of publisher,   (enclosed within brackets)

date of publication,

page number on which article begins, or of specific reference.

For example:

> 5.   Jonathan Ng, "Interpersonal Communication with Asian Businesspeople," *Marketing in Pacific Rim Countries,* ed Christine L. Summakindt (Portland, OR: Dover Books Inc, 2001), p 268.

Second or Third Edition of a Book. If a book is a second or subsequent edition, the words "2nd ed" (or 3rd, etc) should be entered immediately after the book title, as has been done in entry 2.

Article in a Magazine or Journal. The entry should contain

author's name (or authors' names),

title of article (in quotation marks),

title of magazine or journal (in italics),

volume and issue numbers (shown as numerals only, e.g. 17:4),

magazine or journal date,

page number on which article starts, or of specific reference.

For example:

> 6.   Dana Winterton, "Entrepreneurs in a Free Trade Environment," *Business-North,* 14:2, February 2000, p 27.

If the author of a magazine article is not identified, the reference should start with the article title.

Report Written by Yourself or Another Person. The entry should include

author's name (or authors' names, if authors are identified),

title of report (in italics),

report number or identification (if applicable),

name and location of organization issuing report,

date of report,

specific page number (if applicable).

Here is an entry for the formal business report in Chapter 8:

> 7. Dan Rogerson, *Strategy for Remedying the Decrease in Sales at Provo Catalog Order Centers.* Report: L. V. Morton and Associates, Rochester, NY, February 28, 2002.

Excerpt from a Web Page. If the information is presented *only* on the Web page (i.e. there is no printed equivalent), the entry should contain

author's name (if an author is identified),

title of the specific piece of information (within quotation marks),

title of the "document" (in italics)

the date the information was entered (day [numeral], month [spelled out], year [numeral], and

the Web identification <within angle brackets>.

For example:

> 8. J. James Conklin, "How to Write Proposals That Win!" in *TCI 99: The Fourth Annual Technical Communication Institute.* 19 January 1999 <http://www.umanitoba.ca/faculties/con_ed/partners/tci>.

If the information has also been published (in print form), then the entry should contain

the full printed identification (for a book, article, paper, etc.),

the date the information was entered on the Web site (day, month, year), and

the Web identification <within angle brackets>.

For example:

> 9. "Are You Drowning in Email?" in *RGI News*, No. 3, Fall 1998. 2 February 1999 <http://www.rgi-intl.com>.

Technical Paper Presented at a Conference. The entry should contain

author's name (or authors' names),

title of paper (in italics),

name of conference and sponsoring organization,

location of conference,

date of presentation.

For example:

> 10. Marvin Kreston, *Teaching Technical Managers How to Communicate.* Conference on Integrating Business and Education Needs, St. Paul, MN, November 18, 1999.

Letter, Memo, or Email. The entry should have

author's name,

author's identification (employer and location),

form of correspondence (letter, memorandum, email),

addressee's identification (employer and location),

date of letter, memo, or email.

For example:

> 11. Maurice Aubert, Meridian Laboratories, Dallas, TX. Email to Ken Fong, Vancourt Business Systems Inc., Phoenix, AZ, January 22, 2001.

But remember that email is like a conversation (see example 13, below): often there is no documented record of the exchange of information. So, if you plan to refer to an email message, print a copy and keep it on file.

CD-ROM, Floppy Disk, or Magnetic Tape. An electronic storage medium normally is used to store, in a very compact form, lengthy documents and information that have been published elsewhere. Therefore, the source reference must contain information about the original document as well as identification details of the electronic source:

author's name (if an author is identified),

title of excerpt (in quotation marks),

title of publication (in italics),

name of electronic medium (e.g. CD-ROM, disk, magnetic tape), and

city of publication,

name of publisher,   (of the electronic medium; and in brackets)

date of publication.

For example:

> 12. Thomas L. Warren, "Cultural Influences on International Communication" in *ISTC Golden Opportunities Anniversary CD.* CD-ROM (Letchworth, Herts, UK: The Institute of Scientific & Technical Communicators, August 1998).

Speech or Conversation. The entry should be

speaker's name,

speaker's identification (employer and location),

form of communication (speech, conversation, telephone call),

listener's name (individual or group),

listener's identification (employer and/or location),

date of communication.

For example:

13. Juha Nordlund, Nokia Telecommunications OY, Tampere, Finland, speaking at IPCC 98, the IEEE 1998 International Professional Communication Conference, Québec, QB, September 24, 1998.

14. David G. Ainslie, Western Supplies Inc., Seattle, WA, in conversation with Douglas G. Jerome, NOR-ED Distributors, Tucson, AZ, December 13, 2000.

Second Reference to a Document. When a document is referred to more than once, an abbreviated reference containing only the author's surname (or authors' surnames) and new page number can be used for all subsequent entries. If, for example, further references are made to the documents listed earlier as entries 2 and 6, the new entries would be

15. Blicq and Moretto, p 126.

16. Winterton, p 31.

See how Dan Rogerson has done this in the list of references for his formal report, page 149. (Note that the Latin terms *ibid.* and *op. cit.* are not used in modern reports.)

If several documents by the same author are referenced, then the date of publication is included in subsequent entries (to identify which of the author's specific works is being referred to):

17. Carter, 2001, p 147.

Bibliographies

A bibliography is used when a report writer wants to list more documents than are referred to in the report. It may be a comprehensive list of all documents pertaining to the topic being discussed, or it may be limited to the sources that were used to research and conduct the project or study.

The information in a bibliography is almost identical to the information in a list of references, but the entries are presented differently. In a bibliography entry the following rules apply:

- The first-named author's names are reversed, with the surname or family name shown first and the personal name(s) shown second (for example: Blicq, Ron). The second author's names are listed in natural order (e.g. Lisa Moretto).

- The first line of each entry is extended about 0.4 in. (12 mm) or five typewriter spaces to the left of all other lines in the entry.

- The entries are listed in alphabetical order of first-named authors, so that in Figure 11–1 Ainslie appears before Kreston, which appears before Winterton.

- The entries are not preceded by an identification number.

- Within each entry the information is divided into three compartments that are separated by periods:

  Author identification.

  Document or article title.

  Publisher identification.

For example:

> Smithers, Janet, and William Corcoran. "In Search of Quality Measurements." *Technology Newsletter*, 27:15, June 15, 2000.

The documents listed earlier as references are shown rearranged into a bibliography in Figure 11–1. Like a list of references, a bibliography appears at the end of the report narrative, but before the attachments or appendixes.

Note that there is no reference to the Jonathan Ng article in the bibliography, since there is already a reference to the anthology in which the article appeared (under the name of the editor, Christine L. Summakindt). If the anthology had not been included, then the article would have been entered and would have appeared like this:

> Ng, Jonathan. "Interpersonal Communication with Asian Businesspeople." *Marketing in Pacific Rim Countries*, ed Christine L. Summakindt. Portland, OR: Dover Books Inc, 2001.

Footnotes

Footnotes are not recommended for business and technical reports. Their position at the foot of the page not only distracts the reader's eye and interrupts reading continuity, but also creates difficulties when typing the report. Footnotes are better replaced by endnotes (that is, as a list of references at the end of the report), which more conveniently and unobtrusively serve the same purpose.

Figure 11–1   A bibliography.

Bibliography

Ainslie, David G., Western Supplies Inc., Seattle, WA, in conversation with Douglas G. Jerome, NOR-ED Distributors, Tucson, AZ, December 13, 2000.

"Are You Drowning in Email?" *RGI News*, No. 3, Fall 1998. 2 February 1999 <http://www.rgi-intl.com>.

Aubert, Maurice, Meridian Laboratories, Dallas, TX. Email to Ken Fong, Vancourt Business Systems Inc., Phoenix, AZ. January 22, 2001.

Blicq, Ron, and Lisa Moretto. *Writing Reports to Get Results*, 3rd ed. New York: IEEE Press, 2001.

Conklin, J. James. "How to Write Proposals That Win!" *TCI 99: The Fourth Annual Technical Communication Institute*. 19 January 1999 <http://www.umanitoba.ca/faculties/con_ed/partners/tci>.

Gerbrandt, Mavis. *Neural and Fiberoptic Networks*. Los Angeles, CA: Technical Associates Inc., 2000.

Gneiss, Peter L., and others. *Introducing New Technology to the Developing Nations*. New York: Scientific and Technical Press, 1999.

Kreston, Marvin. *Teaching Technical Managers How to Communicate*. Conference on Integrating Business and Education Needs, St. Paul, MN. November 18, 1999.

Rogerson, Dan. *Strategy for Remedying the Decrease in Sales at Provo Catalog Order Centers*. Report: L. V. Morton and Associates, Rochester, NY. February 28, 2002.

Summakindt, Christine L., ed. *Marketing in Pacific Rim Countries*. Portland, OR: Dover Books Inc., 2001.

Winterton, Dana. "Entrepreneurs in a Free Trade Environment." *Business-North*, 14:2, February 2000.

Planning for Reference/Bibliography Entries

When you know there will be references to other documents or sources of information within a proposal or report, we recommend you do some planning to simplify their insertion before you begin typing. To create a reference list manually, follow these steps:

1. Decide whether you will be using a list of references or a bibliography to identify your documentation sources.

2. Create a separate file and label it **References** (or **Refrnces** if your file label is limited to only eight characters). Into it type your full list of references or bibliography entries. Print a hard copy and keep it beside you as you type.

3. Annotate your writing outline to identify where you expect each reference will occur.

4. As you type the proposal or report, key in the appropriate reference number or author identification each time you need to refer to a source document.

Alternatively, you can let your word-processing program's referencing feature create the list automatically as you type your report. When you type information that needs to be source-referenced, key in the reference details at that moment and instruct the word-processing program to store the information until the report is finished. When you have finished the report, the program presents the list to you—either as a reference list or a bibliography, depending on your preference—with all the entries arranged in the proper sequence. The only disadvantage with this process is that it interrupts your writing continuity.

References

1.   *The Chicago Manual of Style,* 14th ed (Chicago, IL: The University of Chicago Press, 1993).

2.   Kate L Turabian, *A Manual for Writers of Term Papers, Theses, and Dissertations,* 6th ed (Chicago, IL: The University of Chicago Press, 1996).

3.   Joseph Gibaldi, *MLA Handbook for Writers of Research Papers,* 5th ed (New York: The Modern Language Association of America, 1999).

CHAPTER 12

Inserting Illustrations into Reports

An illustration can help readers understand more readily a difficult part of a report or a particular point a report writer wants to make. You should integrate text and visuals since some people *need* the visual or graphic description to understand the information or concept. In the same way, you should never present chart after chart without text to interpret the information, since some people *need* the narrative description to understand the information. Because its role is to enhance rather than duplicate the narrative, an illustration must be simple, clear, and useful. A reader should rarely have to turn to the report's words to understand an illustration.

Illustrations appear mostly in longer, more formal reports, such as analyses, feasibility studies, proposals, and investigation or evaluation reports.

To help you select and design the most effective illustration for a given situation, first ask yourself three questions:

1. Which kind of illustration (e.g. table, graph, bar chart, flow diagram, photograph, etc) will best illustrate the particular feature or characteristic I want my readers to comprehend?

2. Will readers be using the illustration simply to gain a visual impression of an aspect being discussed, or will they be expected to extract information from it?

3. Will the illustration be referred to only once, to amplify or explain a point, or will it be referred to several times in the report narrative? (If it will be referred to frequently, its position needs to be carefully considered.)

Some General Guidelines

Number each illustration sequentially, and always refer to it in the report narrative, like this:

> ... in Figure 2 the monthly profits for financial year 2000–2001 are compared with those for the two previous years.

Give every illustration a title:

> Fig. 2. Financial year 2000–2001 profits compared to two previous years.

Sometimes you may want to follow the title with a caption that draws attention to an important point or explains some aspect in more detail. For example (continuing from the Figure 2 title):

> ... the profits for financial year 2000–2001 are compared with those for the two previous years. The curves for the two previous years show a distinct flattening. This flattening is not evident in the 2000–2001 curve.

You must also decide whether an illustration should be placed directly in or beside the report narrative or as an attachment or appendix at the end of the report. For example:

- If the illustration is extremely complex or fills more than one page, insert it as an attachment.

- If readers will need to refer to the illustration as they read the report, place it in the report narrative.

- If an illustration meets both the previous criteria, then insert the complete illustration as an attachment and provide a smaller, much less detailed illustration within the report proper.

Using Computer Software to Produce Graphics

Computer software has taken much of the drudgery out of illustration preparation, yet it must be used with care. You will need to select, from a range of graphs, charts, etc., offered by the graphics software, which will best suit your needs; ideally, select the simplest possible illustration. Occasionally, too, you may have to adapt or modify a graph or chart to ensure that it presents its information effectively. The following section describes various ways to present information graphically and explains what type of information is best suited for each.

Tables

Tables document information in tabular form, such as results of tests, quantities of items manufactured, daily receipts, etc. Unlike many of the illustrations described in this chapter, tables are meant to be examined in detail by the reader, who may want to extract or extrapolate data from them. Consequently, the rules for preparing tables differ from the rules for preparing illustrations such as graphs and charts.

Guidelines for preparing tables are:

1. Keep the table simple, using as few columns as possible.

2. Limit the amount of data by omitting any details readers will not need.

3. Insert a clear, simple, but fully understandable title at the head of each column.

4. Insert a unit of measurement at the head of a column rather than repeat the unit after each entry within the column. (See how this has been done for % in the table in Figure 12–1.)

5. Insert the table number and an informative title, and center them immediately above the table. (*Note:* Table numbers and titles should appear *above* the table, whereas figure numbers and titles should be placed *below* the figure.) This may vary in textbooks.

6. Decide whether the table is to be open (without ruled lines separating the columns,

Figure 12–1   An open table (no lines separate the columns of data).

Table 4

Quality Control Inspection Report: Warrendale Plant

Production Tests: November 1, 2000 — January 31, 2001

Chip No.	No. Chips Manufactured	No. Chips Tested	% of Production Run	No. Chips Failed Test	Failure Rate (%)
AR-17	13 318	480	3.90	9	1.87
CM-20	11 406	300	2.63	2	0.66
FL-06	23 061	460	1.99	3	0.65
RG-14	19 800	375	1.89	1	0.27
RL-08	13 200	260	1.97	5	1.92
RL-21	118 600	1 820	0.69	7	0.38
VX-07	14 087	260	1.85	2	0.52
WR-01*	5 000	330	6.60	12	3.63

*New Product: Production run started December 15, 2000.

as in Figure 12–1 and the attachment to the investigation report on page 81) or closed (with the ruled lines inserted, as in Appendix B on page 152).

7. Ensure that the report narrative tells readers what they should learn from the table, so that its relevance is clear.

Most word-processing and spreadsheet packages can auto-format tables. This function will quickly format your tables so they are attractive and easy to read. Shading rows, highlighting column titles, and inserting bold gridlines are some examples of the formatting used. Explore your software and become familiar with this feature.

Graphs

Graphs offer a simple way to illustrate how one factor affects or is affected by another. They have the particular advantage that the changes they depict can be readily visualized and understood by most readers. For example, graphs can be used to show

- predicted sales for various price structures,
- the extent that electric power consumption increases as the ambient temperature approaches high or low extremes,
- the radar detection distance for aircraft approaching at different altitudes, or
- the life expectancy of a motor operated at varying speeds.

No matter how technical the subject, a graph must be kept simple. The guidelines listed here contribute to this cardinal rule.

1. Limit the number of curves on a graph to three if the curves cross one another, or to four if they do not intersect or there is only a simple intersection. If you have to construct a multiple-curve graph containing five or more curves, construct two graphs rather than one.

2. Differentiate between curves, particularly if they intersect, by assigning them different weights. Make the most important curve a bold line, the next most important a light line, the third curve a series of dashes, and the least important curve a series of dots (see Figure 12–2). Compare this figure with the software-generated graph in Figure 12–3, which shows the curves as a series of short, straight lines. Note, too, that each curve is labelled in Figure 12–3 because the software could not place a different weight on each curve. Unless your office copier can print in color, avoid using color to differentiate between curves.

3. Position the curves so they are reasonably centerd within the frame provided by the graph's axes. If necessary, adjust the starting point of the scale(s) to move an off-center curve to a more central position. See Figures 12–4 and 12–5.

Figure 12–2   A graph with four curves. The most important curve is identified by a bold line.

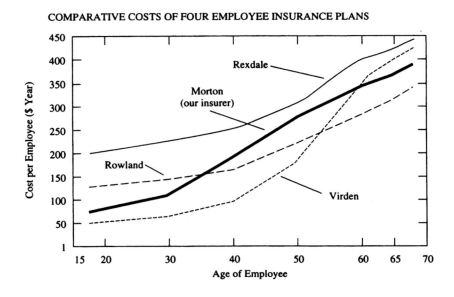

COMPARATIVE COSTS OF FOUR EMPLOYEE INSURANCE PLANS

Figure 12–3   A graph prepared with computer graphics software. The curves are made up of a series of straight lines, rather than smooth curves.

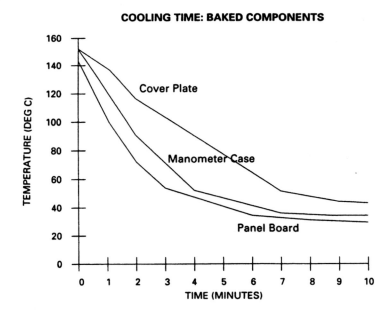

COOLING TIME: BAKED COMPONENTS

4. Select a scale interval for each axis that will help the curves create a visually accurate image (Figure 12–6). An incorrect scale interval may inadvertently cause curves to depict a false impression, as shown in Figure 12–7.

5. Omit all plot points, to provide a clean, uncluttered illustration. The only time that plot points or lines should be present is in a detailed drawing placed in an attachment, from which readers are expected to extract information or examine the graph's construction. (If you are using computer software, you may have to override a default to prevent it from automatically inserting plot points.)

Figure 12–4 An incorrectly centered graph. Although technically accurate, the graph appears unbalanced.

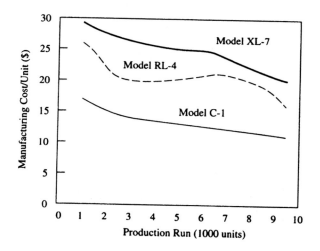

Figure 12–5 A correctly centered graph. The vertical scale starts at 10 rather than at 0, as in Figure 12–4.

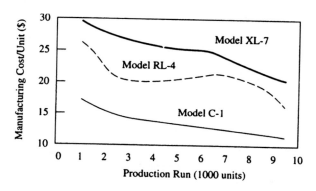

6. Keep all lettering clear, brief, and *horizontal*. The only nonhorizontal lettering should be along the vertical axis, as shown in Figures 12–2 to 12–6. Particularly avoid placing lettering along the slope of a curve.

7. Omit a grid unless you expect your readers will want to extract their own figures from the graph (compare the no-grid graph in Figure 12–5 with the gridded graph

Figure 12–6   Properly balanced scale intervals produce a visually accurate curve.

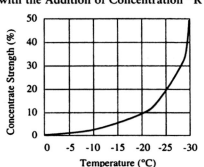

Figure 12–7   The effect of an improperly balanced scale interval. Although these curves are technically accurate, neither creates the same visual impression as the correctly balanced curve in Figure 12–6. The contracted axes over-accentuate the flattening at one end and de-emphasize the flattening at the other end of each curve.

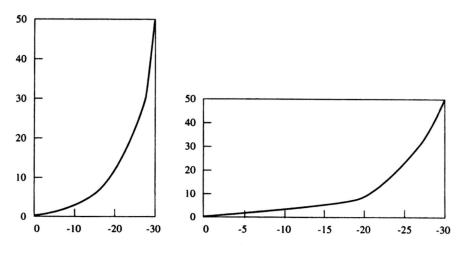

in Figure 12–6). If you are unsure whether to insert grid lines, you can insert an "implied" grid, as in Figure 12–2.

Bar Charts

Whereas graphs have two continuous variables, bar charts have only one. They are simpler to read and understand than graphs and so are particularly useful as illustrations for nonspecialist or lay readers. Normally they provide only a general indication of results, quantity, time, etc, which makes them unsuitable for depicting exact units of measurement; readers cannot extrapolate detailed or exact information from them.

A bar chart offers comparisons using parallel bars of varying lengths to portray weight, growth, cost, life expectancy, etc, of various items. The bars are arranged either vertically or horizontally, depending on the factors being displayed, with the variable function lying along the axis that runs parallel to the bars.

General guidelines for preparing bar charts are:

1. Position the bars so they are spaced one-half to one full bar-width apart. (You may find that the default in some computer software provides even narrower spacing, which is acceptable providing the chart can be clearly understood.)

2. Arrange the bars vertically when you are portraying growth factors, such as weight, quantity, cost, or units produced (see Figure 12–8).

3. Arrange the bars horizontally when you are portraying elapsed time or factors in which time is a significant element (i.e. life expectancy, production time, project length), as shown in Figure 12–9.

4. Shade the bars if you need to make them stand out.

5. If it is important for readers to know the exact total each bar represents, show the totals either immediately above the tops of the bars (if the figures are short enough) or inside the bars (along their length), as shown in Figure 12–10.

6. If the bars are composed of several segments, either identify the segments by various types of shading (and provide a legend beside or below the chart) or, if there is room, identify each segment with a word or two inside the bars (see Figure 12–11). An unusual illustrative technique is to use a picture of a car, person, building, etc, in the size or the comparison (see Figure 12–12). This technique is used more often in magazines and newspapers than in business reports.

Figure 12–8 A bar chart with vertical bars.

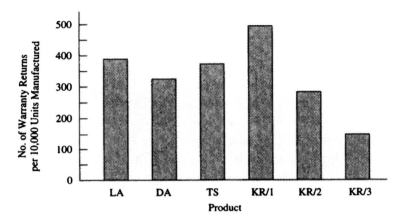

Figure 12–9 A bar chart with horizontal bars.

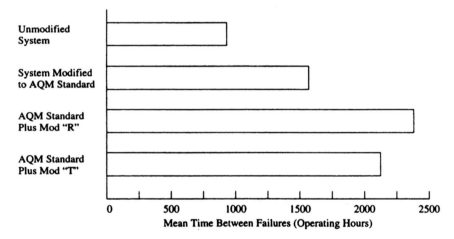

Figure 12–10 Numbers placed above or within bars show exact figures.

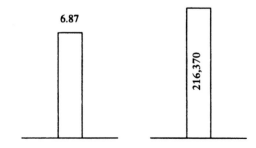

Figure 12–11 Bars can be divided into segments either by shading or by lettering.

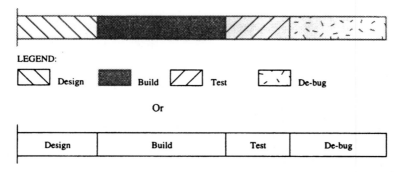

Figure 12–12 A pictorial bar chart. The horizontal axis is broken and shortened between 225 and 775 to permit the long Automobiles bar to be depicted without unbalancing the illustration.

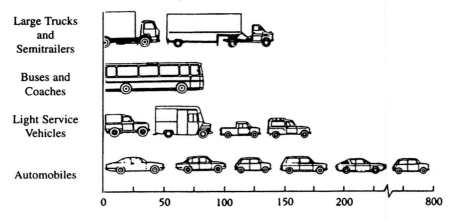

Histograms

A histogram contains some features common to both a graph and a bar chart. It has two continuous variables, but is constructed like a bar chart because there is insufficient data on which to plot a true curve. To show that it has this dual but limited function, the bars are plotted immediately against each other, as in Figure 12–13. Indeed, a line drawn through the tops of the bars would produce a rudimentary curve.

Guidelines for preparing a histogram are similar to those for preparing a graph and a bar chart.

Figure 12–13 A histogram.

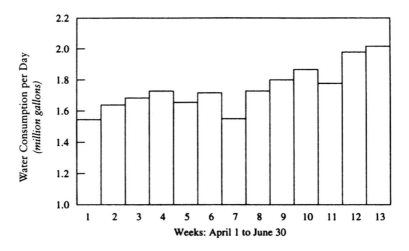

Surface Charts

A surface chart combines the characteristics of both a graph and a histogram. It has two continuous variables, and it is made up of adjoining bars from which the vertical construction lines have been erased (see Figure 12–14). But the curves, instead of being compared as in a graph, are summated; that is, for each vertical "bar," the factors being displayed are added, so that the first curve becomes the base for the second curve, and the second curve becomes the base for the third curve. Thus the uppermost curve represents the total of all the curves added together.

Readers cannot easily extrapolate information from surface charts because direct readings can be extracted only from the lowest and uppermost curves.

Guidelines for preparing a surface chart are:

1. For the lowest curve, choose and draw in the factor that is the most important, represents the largest quantity to be depicted, or offers the most stable (even) curve ("Other" in Figure 12–14).

2. For the second curve, select the next factor and plot it in, using the first curve as the base for each section ("Residential" in Figure 12–14).

3. Repeat the sequence for each additional curve ("Transportation" in Figure 12–14).

4. Shade or crosshatch the curves, preferably making the lowest section the darkest and the uppermost section the lightest.

Figure 12–14 A surface chart.

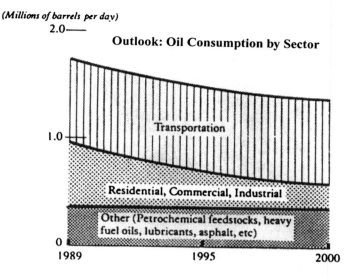

Pie Charts

A pie chart (Figure 12–15) is one of the simplest forms of illustration. By dividing a circle (a "pie") into segments of varying size, we can illustrate such things as market distributions, tax apportionment, and product costs in a readily understandable form. Because it is a simple illustration, only a few guidelines are necessary:

1. Always make the segments of a pie chart add up to 1, 100%, or $1.00 (or a round figure multiple: $100, $1 million).

2. Check that the segments are visually accurate—i.e. that they are in the correct proportions for the quantities they depict.

3. Ensure that one of the dividing lines between segments is vertical, running from the center of the pie to the top (12 o'clock position).

4. If, in addition to the major segments, there are several very small segments to depict, combine them into one segment and label it "Miscellaneous" (or use a more descriptive term). If it is important for readers to know the composition of this segment, provide a list beside the illustration or in a caption below the chart.

The pie chart is one illustration that can often benefit from being depicted three-dimensionally, as shown in the computer-software-produced Figure 12–16.

Figure 12–15 A pie chart.

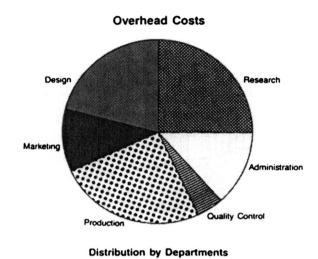

Figure 12–16 A computer-software-produced pie chart, with one segment emphasized by being pulled partly away from the pie.

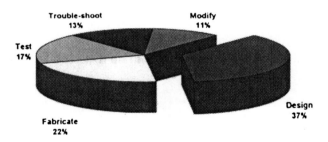

Flowcharts, Site Plans, and Line Diagrams

A flowchart (see Figure 12–17) provides a visual description of a procedure, process, plan, or system. A site plan depicts the more significant features of a building site or small area of a town, whereas a line diagram can encompass anything that needs to be illustrated (e.g. a piece of equipment, hookup of several instruments, layout of an office, etc.); see the manometer/digital recorder test hookup on page 61 and the floor plan in the semi-formal proposal on page 95.

Figure 12–17 A flowchart (also known as a flow diagram).

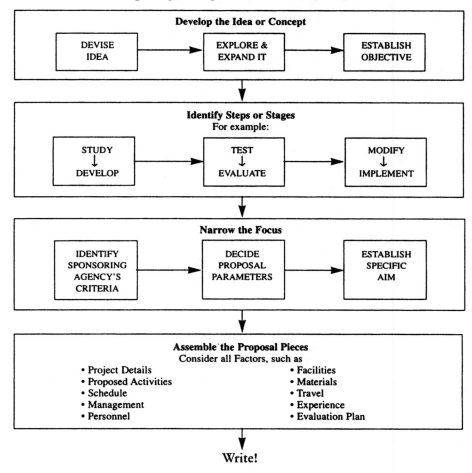

Preparing a Proposal (Pre-Writing Steps)

In all cases they should

- be as simple as possible,
- clarify the accompanying written description,
- contain only the essential elements (which means firmly eliminating unessential elements),
- be easy to follow,
- be readily understood without the written description,
- be drawn in clear black ink, and
- contain neatly lettered, clear but brief explanatory words.

Photographs

Photographs are an ideal, accurate way to show readers either close-up details or "the whole picture," but unfortunately the average office copier does not reproduce them well. To achieve a fine-quality image, photographs have to be carefully prepared for printing by a lithographer, who places a fine dot-screen over them and makes a special plate for printing them onto bond paper with an offset duplicator. Alternatively they can be scanned in by a computer scanner and reprinted by a high-quality color copier.

The Size and Position of Illustrations

The ideal illustration sits beside or immediately above or below the words that refer to it. Unfortunately this is not always easy to achieve, particularly if a report is to be printed on only one side of the paper. If full-page tables and illustrations are inserted into the narrative, they interrupt reading continuity.

Use these guidelines when preparing to key in a report:

- Plan the report's pages before keying them in, even if doing so means having an intermediary draft prepared so you can evaluate how much space each paragraph will require.

- Keep diagrams as simple and as small as possible so there will be room to insert type above, below, or around them.

- Position diagrams so they are adjacent to the paragraphs that refer to them or to the paragraphs that most need illustrative support.

- Beneath every illustration insert the figure number, a brief title, and, possibly, an explanatory caption.

- If a full-page diagram has to be inserted, consider whether it must accompany the report narrative or if it can be placed in an attachment with a small, simple sketch inserted in its place in the body of the report.

- If a full-page diagram is horizontally oriented (i.e. its base is longer than its height), turn it 90° so that it will be read from the right-hand side of the page (see Figure 12–18).

- Check that every illustration is referred to in the report narrative, by quoting either its figure number or its attachment/appendix identification. This applies to illustrations in the body of the report or in the appendix.

Figure 12–18 Full-page horizontal diagrams are turned so they can be read from the right. This should be done even though some words may be inverted when the illustration is viewed from the foot of the page.

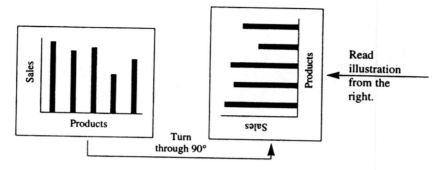

CHAPTER 13

Guidelines for Spelling and Handling Abbreviations and Numbers

Most publishers have a style manual that establishes rules for spelling, capitalization, writing numbers, and so on. Publishers, writers, and editors refer to the style manual as their "bible" and use it whenever a decision on style is necessary (e.g. when they need to know if they should write twelve or 12, mph or m.p.h., calk or caulk). Individual report writers do not have a special style manual to refer to, but they can adopt some of the basic guidelines outlined in this chapter. For more detailed guidelines, we suggest that you refer to the glossaries at the end of *Technically-Write!*[1] or *Communicating at Work*.[2]

Spelling

Select a dictionary published in the US, Canada, or the UK that is large enough to carry most of the words you are likely to use (a 600-page dictionary is ideal), and that has been revised recently (say within the past five years) so that it reflects current spelling practice, such as *Webster's Tenth New Collegiate Dictionary*.[3]

Use the dictionary's spelling rules as *your* rules, so that you will be consistent. Where the dictionary shows alternatives, such as "sulfur; sulphur" and "programed; programmed," generally choose the first-listed spelling because most dictionaries normally list the preferred or more common spelling first. If you decide you prefer the alternative spelling, then underline it with a colored pen so you will be reminded of your

206

preference when you look up the word later. The need to identify an alternative spelling may seem rare, but there will be occasions when you feel that the first-listed spelling is awkward (for example, many people feel "programed" is an unnatural spelling, and choose to write "programmed").

A useful book that discusses style and lists many "problem" words is the *United States Government Printing Office Style Manual*[4] (it lists nearly 20 000 terms). For business and technical report writers, *Communicating at Work* and *Technically-Write!* list many "awkward to handle" words and expressions in their glossaries.

Abbreviations

You may abbreviate any term you wish, particularly if it is a lengthy term that will be used frequently in a report. These are the main guidelines:

1. Indicate to the reader what your abbreviation means by writing it in full the first time and then showing the abbreviated form in brackets beside it:

    > The digital reference number (drf) we applied to the first set .... In the second set, the drf was determined by ....

2. Avoid forming your own abbreviation when another abbreviation already exists and is commonly used.

3. Use lowercase letters, unless the abbreviation is formed from a proper noun such as a person's or a company name:

foot	ft
metre	m
average	avg
ampere	A (from the name Ampere)

4. Omit all punctuation, unless the abbreviated term forms another word:

absolute	abs
approximately	approx
inch	in.
number	no.

5. Omit the *s* from an abbreviation of a plural quantity:

metres	m
numbers	no.
kilograms	kg
hours	hr

Be aware that there are some exceptions to these guidelines, caused by nonstandard terms being adopted through general usage. For example, although no. is the correct abbreviation for number, you are much more likely to see it used as No. (and sometimes even as #, which is definitely nonstandard). You will not be wrong if you choose to use No., but be consistent: whichever term you use, use it all the time. For information on technical abbreviations, refer to the glossary in *Technically-Write!*

Numbers

Publishers' style manuals also reflect standard usage for writing numbers when they appear as part of a sentence or paragraph. The general rule is from one to nine, spell the number. For 10 and higher, use numerals. But there are exceptions to this rule, and sometimes you may have to decide whether the rule or the exception takes precedence.

1. Spell out the number if it is
    *   the first word in a sentence,
    *   a large generalization (as in "... about eight thousand ..."), or
    *   a fraction that is less than one (as in "... only one third of the participants ...").

2. Use numerals if the number
    *   is part of a series of quoted numbers,
    *   is a year, date, time, age (of a person), percentage, or sum of money,
    *   is part of a unit of measurement (as in 28 kg),
    *   is specific technical data, such as a dimension, tolerance, temperature, or result of a test,
    *   contains a decimal or a fraction (as in 6½ and 3.25), or
    *   refers to a chapter, figure (illustration), or page (as in page 138).

Also, insert a "zero" at the start of any decimal that is less than one (unity), as in 0.25 and 0.0056. If the decimal is faded on the printed copy, the reader will know it should be there because there is a space between the "0" and the first digit.

In North America, we often insert a comma (or commas) inside a number with five or more digits: 27,384. In Europe, the comma is replaced by a space: 27 384. This is done to avoid confusion in countries, such as France and Sweden, where the comma is used instead of a decimal point.

Metric (SI) Units

As metrication becomes more firmly established, report writers—and particularly those who write technical reports—have to know the rules for writing metric (SI) units. These were defined originally by the eleventh (1960) Conférence Générale des Poids et

Mesures (CGPM) in Paris, France, and have been only slightly revised in the intervening years. The current guidelines for writing SI symbols are:

1. Use upright (*never italic*) type.

2. Use lowercase letters, except when a symbol is derived from a person's name (as in V for volts, which is derived from Volta).

3. Insert a space between the numeral and the first letter of the symbol (as in 38 kHz), but *no* space between the symbols themselves (e.g. there is no space between k and Hz).

4. Omit the s from all plurals (as in 221 km), but do not insert a period at the end of the symbol (except when the symbol is the last word in a sentence).

5. Insert an oblique stroke for the word per (as in km/hr), and a dot at midletter height to show that two symbols are multiplied (as in N·m, for Newton meter).

For more information, refer to the glossary in *Technically-Write!*

References

1. Ron Blicq and Lisa Moretto, *Technically-Write!* 5th ed (Toronto, ON: Prentice-Hall, 1998).

2. Ron Blicq, *Communicating at Work*, 3rd ed (Toronto, ON: Prentice-Hall, 2001).

3. *Webster's Tenth New Collegiate Dictionary* (Springfield, MA: Merriam-Webster, 1998).

4. *United States Government Printing Office Style Manual*, Superintendent of Documents, Washington DC, 20402.

CHAPTER 14

The Report Writing Process

This chapter provides a simple overview of the five main steps in the writing process. Each step varies in length and, depending on the individual and the particular project, the writing process varies in length and complexity. The steps are:

1. Preparing to Write

2. Organizing the Information

3. Writing the Words (Draft)

4. Editing the Information

5. Revising the Text

If you skip a step you are risking the quality of your finished document.

Preparing to Write

After you have gathered all your information and before you start writing, there are several things you must consider. Chapter 10 discusses the importance of knowing who you are writing to and asking yourself some questions about your audience. Chapter 2 describes the difference between a "tell" message and a "sell" message, and how you must decide right away which type of message you are writing.

You should also consider how the information will be presented on the page. Chapter 9 introduces some simple information design techniques. Most word processors have default settings and templates we can use as starting points for page layout. Often, though, the defaults and templates are either bland and unappealing or very restrictive. After considering your audience and the type of information you are conveying, you may want to adjust some of the settings. Use the online help system to find out how to do it.

Margin Width. The default normally positions the lines so the margins are equal-sized on both sides, left and right, and top and bottom. If you are preparing a report that is to be bound within a folder, you may want to instruct the computer to set a wider margin on the left side of odd-numbered pages and on the right side of even-numbered pages to allow for the binding edge.

Page Number Position. Most software programs automatically position the page numbers at the foot of the page, level with the right-hand margin and approximately three lines below the last line of text. If you prefer the page numbers to be centered you can easily change the default. Sometimes a company logo or confidentiality statement is required on the bottom right corner of the page, which means the page number needs to be placed elsewhere.

Line Spacing. The default setting is usually single spacing between lines of text. If you are writing a first draft of a document and are sending it to others for comments, 1.5 or double spacing between lines is useful because it leaves extra room for people to write in their comments.

Word Hyphenation. You can choose whether words that extend beyond the length of a line are to be hyphenated or carried in full to the start of the next line. You can also choose whether you or the computer will decide where to hyphenate a word. If you are using highly technical, specialized jargon, however, the computer may not recognize the word and the result may be an oddly placed hyphen.

The Number of Lines between Paragraphs. Normally one blank line is used for short, single-spaced reports. For formal reports and proposals, one-and-a-half or two blank lines should be left between paragraphs to create a more open effect. (Two blank lines are always inserted between paragraphs when the whole report is double-spaced.)

Whether to Indent the First Line of Each Paragraph. First lines are not indented in modern correspondence and short reports. In long reports, the same holds true, although there are variations. Whichever you do, be consistent within the same document.

Subparagraph Indents. Properly executed subparagraph indentation provides a visual clue to help your readers see how you are subordinating your ideas. Note how these subparagraphs are indented: each is moved to the right as a complete block of information. Subsubparagraphs can be treated in exactly the same way, except that they are indented a double distance to the right.

Types of Headings. Just as with subparagraphs, headings provide a visual clue to the importance of the information. Software programs provide several type fonts and an almost limitless range of type sizes to choose from (type size is measured in *points*). Use bold letters and a variety of type sizes so that the relative importance of each heading is *visible*. For example, using the Univers type font:

1. # Major Headings in 18 point Univers
2. ## Secondary Headings in 14 point Univers
3. ### Introductory Headings in 12 point Univers

We recommend that you select a simple, well-known type font rather than an exotic one. Then use that font throughout your report, choosing one point size for the text and others for the headings. A major decision when selecting a font is whether you want *serif* or *sans serif* type (serif type has curls on the letter extensions that rise above and drop below the printing line, as in b, d, f, g, h, j, etc). Here are two examples of popular fonts:

Times New Roman in 12 point type size *(serif)*

Univers in 12 point type size *(sans serif)*

Research shows serif fonts are easier to read. We use serif fonts for longer, more technical information and sans serif fonts if we write a one- or two-page letter or memo.

Organizing the Information

At some point in their education, most people were taught how to develop an outline. However, reports can become generic and meaningless if the writer feels constrained by an outline. Often writers end up stuffing information into a structure without considering the content. This section explains how to develop an outline based on the information you have and the audience you are writing to.

You need an outline—even a very basic one—to help keep the design of the report continually before you. You can prepare your outline on paper or you can type it directly into the computer. Some software programs have outlining features to help you.

Step 1. Create a file and assign it a descriptive name of your choice (for example, PRE-OUTL for preliminary outline). Into this file type a list of headings, double-spaced one above the other, each describing a topic you feel should be covered in your report. For example:

Width of shaft

Elevator sizes

Problem with access

Initial failure (May)

Current limitations

... etc

The key to preparing this list is *not* to organize it as you go along. Type whatever topic comes to mind first. And then type another ... and another ... and another. What you will be doing is "brainstorming" ideas into the computer. The sequence in which they appear, their importance, or even their relevance should not be considered at this point. That will come later, when you start arranging the information into a coherent outline.

When you have exhausted all likely topics, save the file. Return to the start of the file and for each topic you listed ask yourself two questions:

- Is the topic really relevant?

- What subtopics does it generate?

If you decide a heading is interesting but not truly relevant, cut and paste it to the bottom of the file. You can review it later to see if it belongs. If the topic generates additional ideas, type them into the file immediately after the topic entry, and indent them to the right to show they are subordinate topics. For example:

Width of shaft

Elevator sizes

    Regular

    Freight

    Executive

Problem with access

    First evidence—November

    Comments—safety inspector

    Meeting—February

Initial failure (May)

... etc

When you feel you have developed a comprehensive list of topics (even though they are in random order), save the file and print out a hard copy to work with.

Step 2. You now have to take the random list and assign each topic to one of the writing compartments of the report. Turn to earlier chapters of this book and decide which type of report you will be writing. Find the pyramid illustration containing the writing plan for the report, and identify the writing compartment names. Give each a one-letter label, such as I for Introduction, A for Approach, F for Findings, and C for Conclusions. On the hard copy, write an appropriate one-letter label either in front of or behind every entry in the list. For example, after "Width of shaft" you might write "I" if you want to provide details of the shaft's size in the Introduction to your report or "F" if you feel the width of the shaft is one of the report's Findings. (With experience you may prefer to do this annotating on screen, without using hard copy.)

Step 3. Now you can form your writing outline. If you originally type the topics in a roughly coherent order—and this sometimes occurs because one topic often generates the next logical topic—you may be able to rearrange them on screen. More often, however, you will find there are so many changes that it is faster to prepare a completely new list.

Create a new file called FIN-OUTL and type the appropriate labels of the writing compartments into it, such as those for an investigation report (see Figure 6–1 on page 70), so that they appear as a series of main headings:

Summary

Introduction

Investigation

    Approach

    Findings

    Suggestions

    Evaluation

Conclusions

Recommendations

Attachments

Under each heading type in the list of topics you previously annotated with that heading's identification letter and arrange them in a logical order. If there are a lot of topics you may have to do this in two steps, first typing the topics in random order and then rearranging them on screen. When you have finished, your writing outline will be ready and you can print a hard copy to use as a guide as you write your report.

If, as you write the report, you find you need to modify the outline or add a section to it, access the FIN-OUTL file, type the alterations, and print a new copy.

Writing the Words (Draft)

Many professional people—all specialists in their fields—have a problem getting their first words onto paper. If this happens to you, the "Getting Started" section in Chapter 2 will help. But to overcome the "writer's block" syndrome we also suggest breaking the report up into manageable sections:

- Rather than consider the report as one, long, single document to be typed from start to finish, think of it as a group of small, self-contained sections. Often we can even divide the sections into separate files.

- Once you have your outline divided into topics, scan it and identify which topics are most interesting, and start to write about them first. (See Dan Rogerson's approach on page 154.)

- Never start at the beginning. The task will seem too long and impossible and you may fall into the trap of reverting to the "Climactic Method" of writing.

- Avoid trying to type a perfect first draft. Recognize that no one—not even a professional writer—expects to produce instantly usable words and sentences that require no further polishing.

- Type with a minimum of editing (save the editing until later). Try to achieve a momentum that will carry the writing along. If you can't seem to find the exact word you are looking for, write a suitable word and put brackets around it so you know when you edit it that you should consider other words. The same is true for spelling. You can correct that later. Don't interrupt the flow.

If you treat a report as one long, coherent document and start writing at the beginning, you are going to be inhibited by the immensity of the task before you even type one word. But if you consider each section of the report as a minireport complete in itself—with an introductory section, a development in the middle, and a concluding section—the task will not seem nearly so forbidding. As well, you will be constructing a report that is cohesive from start to finish and coherent within each section.

Start by writing about a topic with which you are particularly familiar so that your knowledge of and interest in the topic will help you form the initial words, even if they are not quite what you planned to write. Say to yourself:

> "The first few sentences I write will not be nearly as good as I would like, but I will leave them there, on disk, without attempting to revise them now. I'll look at them again later, when I have a better handle on the overall approach I have taken in other parts."

Later on, when you do go back to revise them, you will find the correct words form in your mind much more readily.

Writing is a creative process that demands concentration and continuity. After writing several sentences or paragraphs without interruption, a writer builds a momentum. At such times the creative process is working to its utmost and any interruption can destroy it. If the writer continually stops to revise what is on the screen, the momentum stops.

For that reason you should type a complete section of a report before you attempt to edit it, either on screen or on hard copy. If you suddenly realize how to correct a paragraph that previously dissatisfied you, do not go back to it. Instead, when you reach a comfortable break in what you are currently writing, type notes directly onto the screen describing what you want to do, or totally rewrite the paragraph right where you are. (You can insert a row of asterisks above and below the new sentences, so you can readily identify them.) When you are in the editing stage the revision can easily be copied into the correct place in the file.

When you have completed writing your first section, you have to decide whether to review and edit it right away before you start writing the next section or to continue writing the remaining sections. There is no preferred method; both are widely used. If you feel you have momentum and want to continue writing, you may find it useful to print a hard copy so that you can glance over what you have already written, particularly if you need to correlate the information in one section with the information in another.

Editing the Information

You can edit (read and revise) a document in two ways:

1. You can do all the work directly on the screen.

2. You can print a hard copy and make your revisions on the printed page first and then transfer them to the electronic copy.

Most people decide to do the first editing online (directly on the screen); however, we recommend you do the major editing on a printed hard copy. It is too difficult to see the entire document and page layout on screen, plus your eyes will tire more quickly, which makes it easy to overlook punctuation, grammar, and structural problems.

Initial Proofreading

The first rule of editing is *not* to pick up a pencil or pen right away and start correcting the words as you read them. There are two preliminary steps you should take:

1. Take a break from the document so you can return to it with a fresh eye. This may be an hour, several hours, or a day. If you try correcting sentences and paragraphs immediately after you have finished writing them, you may still have in your mind what you intended to write and not really see what you actually wrote. You will still be influenced by the enthusiasm and momentum of the moment and so might miss ambiguities, factual errors, and awkward sentence constructions.

2. The first time you read a section (or the whole report, if it is short), read all the way through without a pen or pencil in your hand. Your intent should be to view your work the way your readers will view it: as a continuous document that they will read without stopping to make changes. This way, you will be able to check its overall continuity before you become too familiar with the individual parts.

Detailed Editing

Most people consider editing to mean checking that one's work is written well and has no grammatical or typographical errors. This is true, but it is only part of the picture. Proper editing means making a word-by-word check of a document to determine

- its appropriateness,
- its coherence, completeness, correctness, and conciseness, and
- the quality of the writing.

An experienced editor can check all these factors concurrently. We—as inexperienced editors—would be wise to examine them separately.

Checking for Appropriateness. This can be done during the initial read-through. It means stopping first to ask yourself the questions previously discussed in Chapter 10 (see pages 166–179):

1. Who is my reader?
2. What is the purpose of my report?
3. Do I want to be informative or persuasive?

Keep the answers to these three questions continually in mind as you read. Check all the time that the tone and technical level are correct for the intended readers and that you are providing the information they need—not too many details, yet not too few. If you are still unsure about the appropriateness of your writing, there is a second check you can make later (see "Obtaining an Objective Opinion" on pages 220–221).

Checking for Coherence and Completeness. These two factors can also be examined during the initial read-through. Checking for coherence means ensuring that there are logical connections between the different parts of the report, not just within the section you are currently editing. Checking for completeness means ensuring that *all* the information the reader needs has been transmitted. If you prepared a comprehensive outline and stuck to it as you wrote, then you can be reasonably sure that what you have written is complete.

Checking for Correctness. Here you have to examine facts and figures to ensure they have been accurately written. It means meticulously checking every detail in your report, particularly quantities, measurements, and times. When proofreading, especially on screen, we tend to pay attention to the words and *assume* the numbers are correct. Yet it is very easy to transpose a number (for example, to type 7596 when the correct number is 7956) and then not notice the error during proofreading. This is particularly true of numbers buried in a sentence, like this:

A check of the 7596 samples taken at the test site showed that 247 (3.1%) were contaminated.

Because we are checking the readability of the words, the numbers seem to fit in with the flow of the sentence. Yet a reader checking the calculation would discover that 3.1% of 7596 is 235 and would not know which of the three numbers in the sentence is correct. This does not mean you have to recalculate everything during the editing stage; it simply means going back to the source of your figures and checking that you typed them accurately.

Checking for Conciseness. As you proofread you should continually evaluate whether you have presented your information succinctly. There are two factors to consider:

1. There is a tendency to overwrite when typing. We think and type at different speeds. Some people can type faster than they think and others think faster than they can type. In either case, the result is an unorganized flow of text, which is often longer than it needs to be. Because the entire document is not visible on our computer screens we may duplicate information or be too wordy. The only way to catch this is to proofread very carefully.

2. There is a tendency to use clichés and words of low information content (i.e. words that add no value or meaning to the sentence). As Chapter 10 describes under the heading "Avoid Clutter" (see pages 174–177), it is a rare writer who does not occasionally use an expression that sounds nice but adds little to a sentence.

If wordiness is a problem—especially a trend to use big words—you can purchase a software program that will count the size of each word you have typed. It then provides you with a readout that compares your average word size against a "wordiness" scale. Several programs are available, and more are coming onto the market regularly.

Checking for Good Language. This is probably the most difficult factor to assess in your own writing. You should automatically check that you have used good grammar and proper punctuation. You should also check that you have used a definite, informative, readable style in which the active voice occurs more than the passive voice (see pages 172–174), and the subordination of ideas is readily apparent. For more information on these aspects, refer to a standard handbook of English (for example, *The Prentice Hall Handbook for Writers*).

Checking for Spelling and Typographical Errors. Although you have already passed your report through a spell-check program, you still need to proofread it for typographical and spelling errors that the program did not pick up, such as omitting the "d" from the end of "formalized" (which the program would recognize as "formalize" and not flag as an error) or accidentally typing "continual" and "accept" when you meant to write "continuous" and "except." Such proofreading calls for word-by-word scrutiny and is best done as a separate check during which you search *solely* for typos. It is also difficult to do thoroughly on screen.

Take a clean sheet of paper or a ruler and slide it slowly down the hard copy one line at a time. Pause as each line becomes visible and read it slowly and carefully, one word at a time, keeping the next line covered so that you will not be tempted to skip along quickly and start reading it. At the same time check that the commas, colons, semicolons, and periods have been inserted correctly. This line-by-line scrutiny may be slow, but it is effective because you examine individual words without being influenced by complete thoughts. Remember, your report is a reflection of you. You are creating an impression of yourself. Careless proofreading conveys an image of poor work and may affect your credibility and reputation.

Mark each error clearly. Draw a bold circle around the error so that you will notice it when you refer to the page as you make changes on the screen later. Avoid the practice of proofreading beside the computer and of making each correction as you find it. You will lose the dedicated concentration you need for proofreading. For added visibility, use a colored pen or pencil; red or green is ideal.

Editing Checklist. Here is a suggested checklist, posed as a series of questions that you need to answer about your writing:

Is the focus right?

- Have I directed the information to the primary reader?
- Have I summarized the key point(s) in an opening statement?
- Are the important points clearly visible?
- Will the primary reader be able to read all the way through without becoming lost?
- Have I considered secondary readers who also may read the document? Will they understand it?

Is the information correct?

- Is it accurate?
- Is it complete?
- Is all of it relevant (for the particular reader)?
- Have I checked all numerals and cross-references?

Is my language good?

- Is my writing clear and unambiguous?
- Have I eliminated wordy and LIC expressions?
- Have I used the active voice wherever possible?
- Have I used personal pronouns (where appropriate)?

- Have I checked for spelling errors and typos (both using a spell-checker and physically, line by line)?

Have I kept my letter, report, or proposal as short as possible, yet covered the topic in sufficient depth? *Does it meet the readers' needs?*

- Would I want to receive what I have written?

- What reaction will it evoke from the intended reader (the reader who will act, react, approve, or decide what action to take)?

- Is that the reaction I want?

Revising the Text

Doing a Second (or Subsequent) Edit

When all the improvements and corrections have been made, you are ready to do your second edit. This time you have to check only three factors:

1. That the report reads smoothly and coherently and says what you want it to say to the readers you have in mind.

2. That *all* the changes you identified during the first edit have been made.

3. That you have not inadvertently created further spelling or typographical errors while making the corrections.

We suggest you read the report twice: once for readability (item 1), and once for accuracy (items 2 and 3). This time, however, you do not have to make a line-by-line check of the whole report, as you did previously. You can limit your check to those areas where you made changes.

How long should you continue reading the report and making changes? Continue until you feel the report is an effective conveyor of information, keeping these guidelines in mind:

- If the report is going to an important client and a major contract or project depends on how it is received, then you should spend considerable time reading and polishing it.

- If the report is for in-house use and is fairly routine, then you probably need to edit it only once.

Reports that fall between these guidelines will probably need two or three edits.

Obtaining an Objective Opinion

One of the problems of editing your own work is the ability to view it objectively. Often, familiarity with the subject can blind you to your report's deficiencies. Consequently, when a report is particularly important, ask a disinterested person to read it and give you an objective opinion of how well it achieves its purpose. The reader will probably have to be someone who works where you do, but it should be a person who is no more familiar with your project than the intended audience will be and who can view what you have written without bias. It should also be someone who will give you an *honest* opinion of your report: we all like politely phrased words complimenting us on our writing prowess, but when looking for constructive criticism polite words are not much help.

When asking a reviewer to give you an opinion, explain

- why the report needs to be reviewed,

- who the ultimate reader will be,

- what impact or effect you want the report to have on the reader,

- what aspects of the report particularly need the reviewer's attention (you don't want the reviewer to think he or she is being asked to proofread the report, when what you really want is an opinion on the report's persuasiveness or tone), and

- how soon you need the reviewer's comments.

Ideally, write what you want the reviewer to do on a separate sheet and clip it to the front of the report. Ask the reviewer to write rather than tell you his or her opinion and suggestions, either on the sheet of paper or on the report itself. Be specific. If you prefer the comments to be written on the printed version of the report, say so, otherwise the reviewer may email you a list of comments or add the comments into an electronic version. A spoken statement such as "Sounds good to me, though you had better fix up a vague bit on page seven" is itself too vague to be of much help. And when you receive the reviewer's comments, welcome them even if you disagree with them; try not to be defensive if the reviewer says things about your writing that make you feel uncomfortable!

CHAPTER 15

Guidelines for Working with a Report Production Team

Sometimes you will work independently, writing, typing, and printing (on an office printer) a report entirely on your own. At other times you may work as part of a report production team, co-writing a report or proposal with other engineering writers, probably working with illustrators and possibly editors and a printer. If you are working alone, you are solely responsible for the quality of your product, which includes the appearance of the finished work as well as the correctness of the words you write.

The impression that readers gain of a report writer and the company or organization he or she works for is influenced directly by what they see and read. Words poorly centered on the title page, typing errors, misspelled words, unevenly positioned page numbers, and grammatically incorrect sentences create an image of a sloppy worker employed by an organization that produces a low-quality product or service. Good language, however, and crisp, clear typing neatly positioned on every page convey the image of a confident report writer employed by a highly professional organization.

The impression you convey is equally important when you write as part of a team, because then everyone contributes to and is part of the image that readers perceive. Writing a cooperative report means working with the team members: co-authors, illustrators, editors, and printer. The guidelines presented in this chapter suggest ways for achieving a harmonious atmosphere and producing a high-quality, jointly written product.

Working with Management

If, as sometimes happens, the final draft of a report you have written does not meet management's expectations, you may be asked to make more changes than you expect. You can avoid the frustration this creates by going to your manager or immediate supervisor and asking for guidelines *before* you start writing. Tell your manager that you need to know

- who the primary reader is, who the secondary readers are likely to be, and who is most likely to use the information you supply or take action as a result of your report,

- if you can use the pyramid technique (main message up front) for *all* your reports, regardless of whether they contain good or bad news,

- if you can use the first person (I or we) in your reports, and

- if you can use the more emphatic active voice, rather than the dull, less interesting passive voice.

Some managers may not be aware of these techniques and you may have to convince them of their merits. Use this book as evidence of how effective well-written reports can be.

Finally, when you submit your draft report for evaluation, send out a copy that is absolutely clean (i.e. has no pencilled or ink alterations). A marked-up copy *invites* the reviewer to suggest more changes!

Working with Other Writers

To its readers, a collaboratively written report or proposal should appear seamless, as though only one person has written it; that is, its readers should not be able to detect differences in style and approach. This demands significant cooperation between every person writing sections of the report.

Before anyone starts writing, you should meet as a team and establish specific guidelines that will apply to every writer. Summarize the outcome of the meeting and circulate a copy to each contributor. Here are some suggestions:

- Appoint one writer to be coordinating writer/editor. Agree that this person will do the final editing and will ensure that the parts fit together as a cohesive whole.

- Ensure that everyone understands

  - the purpose of the report or proposal and what it is to achieve,

   – who the primary reader is (or primary readers are), what they currently know about the topic, and what their expectations are (the primary reader is the person or persons who will approve a proposal or make a decision based on a report's findings and recommendations), and

   – who the secondary readers are, why they will be seeing the report or proposal, and what they know about the topic.

- Determine who writes what and roughly how long each section is to be. Decide if one person or the group will develop the outline and structure of the report or proposal.

- Establish a schedule: include the dates when first drafts are due, reviews are to be complete, and changes are to be incorporated.

- Decide which word-processing software will be used, and establish format parameters such as the number of lines per page, line length, type font, size and style of headings and subheadings, and whether the first line of each paragraph is to be indented (see the section on preparing to write in Chapter 14). This will reduce the number of editing changes that have to be made later.

- Decide on language-handling guidelines such as

   – acronyms and abbreviations for common terms,

   – whether the first person is to be used (for a collaborative work, almost always "we"), and

   – whether the active voice rather than the passive voice is to be used whenever possible.

- Decide on a common writing plan, or "shape," for each section. For example: establish that each section will open with a Summary Statement that summarizes what the section will cover; then organize the remaining parts into Background Information, Details, and a Conclusion that sums up the key points. (This is not meant to inhibit the creativity of individual writers but to give them a general framework within which to present their information.)

- Plan to meet regularly (but briefly) to report progress and identify problems that may be affecting the project.

- Develop some guidelines for using email to communicate with other team members.

   Identifying these factors early in the report or proposal writing project will ensure that everyone understands the approach and that everyone is working toward a common goal. This will help overcome problems that may arise later.

Using Email to Communicate with Others

Email definitely improves the communication among project team members. However, if it is overused or used without discretion it may be counterproductive.

There are no established guidelines for the proper way to use email, but we can give you some suggestions that will help you be a good email communicator. Remember that just because email is immediate and somewhat less formal than other methods of communication, it doesn't mean you can

• write snippets of disconnected information,

• write incorrectly constructed sentences,

• forget about using proper punctuation,

• ignore misspelled words, or

• be abrupt or impolite.

Remember that your email messages are creating an image of you and your company or organization.

Although it may seem as if you are simply sending the message to your identified distribution list, email is not a confidential medium. The message you intend for only your co-workers can easily be forwarded to upper management or even outside the company to a client. If you write sloppy, incoherent email messages you are presenting a sloppy, incoherent image of yourself and your work.

Write "Pyramid Style." You can use the pyramid method for writing email messages, just as you do for letters, reports, and proposals:

1. Start with what you most want your reader to know and, if appropriate, what action you want the reader to take.

2. Follow with any background information the reader may need to understand the reason for your message, and provide details about any point that may need further explanation.

3. Check that each message contains *only* the information your reader will need to respond—and no more. That is, take care to separate the essential *need to know* information from the less important *nice to know* details.

Remember that busy readers who receive many messages want them to be concise yet complete. Feed their needs. Email software provides the writer with the opportunity to write a high-level summary of the information and then attach a file containing the complete details. Too often people send an attachment and never use the message part for any more than to write "Here's the file I told you I would send you." People will appreciate it if you tell them, very briefly, what is in the attachment and why they need that information. Then they can open the attachment when they need it. It saves time that is often very precious.

If you are writing to multiple readers, consider sending *two* messages rather than a single all-embracing message. Write

1. a short summary, which you send to readers who are interested only in the main event and the result, and

2. a detailed message, to readers who need all the details.

Proofread with Care. Proofread email *very* carefully: the informality of the medium and the speed with which you can create and answer messages can invite carelessness. It is very difficult to find errors online. We are too familiar with the message we have written and our eyes may skip over poorly constructed sentences or low information content expressions. We suggest printing your messages and proofreading them on hard copy. This is especially true for important or sensitive messages.

Be Considerate of Others. Remember that email is not a good medium for conveying confidential information, or for making uncomplimentary remarks about other people. Email messages can too easily be forwarded or copied to other readers, and then you have no control over who else may see what you have written. Be just as professional as you are when writing regular letters and memorandums.

Similarly, be just as sensitive when deciding to forward a message to another person. Be sure that the original sender would want his or her message distributed to a wider audience.

Emailing information and reports back and forth speeds up the writing process, but remember it isn't the only way to communicate. The telephone is more personal because the person can hear your voice and the intonation can express urgency, excitement, or gratitude. A fax machine works well if you want to annotate a diagram quickly and easily without worrying about technology. And don't forget, there are times when it is more effective to deliver your message in person to someone in the same building.

Working with Illustrators, Draftspersons, and Graphic Artists

Good communication is extremely important between a report writer and an illustrator, draftsperson, or graphic artist (here, we will use the one term: illustrator). If the illustrations in your report are to complement the words you have written, your illustrator needs to know something about the topic, the purpose of the report, who the reader(s) will be, and what aspects need to be emphasized.

There are five guidelines for achieving effective communication between yourself and your illustrator:

- Explain the purpose of the report and of each illustration.

- Discuss how each illustration is to support your words (give the illustrator a draft copy of your report to read) and describe what parts are most important. If possible, sketch each illustration as best you can and then give the sketch to the illustrator so he or she can visualize what you have in mind.

- Provide accurate vertical and horizontal dimensions for each illustration. If the illustration is to be preceded or followed by typing, allow sufficient space between the illustration and the text, so that the page does not look crowded. (If your illustrator is using graphics software, sizing adjustments will be much easier to achieve.)

- Be sure to allow the illustrator plenty of time. Quote an illustration completion date (preferably in writing), and obtain the illustrator's assurance that the date can be met. Never say you want your illustrations ASAP (as soon as possible).

- Discuss oversize drawings and determine how much they will be reduced photographically, so that the illustrator will know not to make construction lines too light.

Working with a Printer

Most business and technical reports are printed in-house or by a local "quick copy" service. In both cases the reports are duplicated on an office copier if only a few copies are required or on a high-speed duplicator if more than, say, 50 copies are required. (A report is rarely taken to a professional printer, except for corporate annual reports.) If your report is printed in-house and the production run is short, you can make the copies on an office copier. If your organization has its own print shop or if you use a copy service, then you should discuss the job with the person who does the printing.

Guidelines for working with a printer are:

- Discuss the report with the printer before the final draft and illustrations are done. Find out what equipment the printer has and if there are any special requirements or limitations. If the printer requests an electronic copy, ask what format or software programs are acceptable. Mention the date you plan to bring the job in, and ask how long the printing will take.

- If the report is large or many copies are required, visit several printers and ask for cost estimates. Be sure to give the same requirements to each printer (e.g. the number of pages and number of copies, how many photographs are being used, and whether the printer is to collate and bind the report).

- When you take the job in for printing, write clear, complete instructions to the printer and clip or staple them to the job. Your instructions should include
  - number of copies required,
  - color of ink to be used,
  - weight of paper (e.g. 20 lb bond),
  - size of paper (e.g. 8.5 x 11 in.),
  - whether the report is to be printed on one or both sides of the paper,
  - where photographs and drawings are to be inserted,
  - whether the job is to be collated and bound, and the type of binding required,
  - any special instructions, and
  - the date you require the printed report.

- If the report is at all complex, make a mock-up showing how the finished product should appear. Use the correct number of blank sheets of paper, fold them once, and staple them together to form a booklet. Open up the booklet and write a descriptive word on each page to show what should be printed there. If a page is to be left blank, write BLANK PAGE on it. The printer will use the booklet to determine what to print on each page and to assemble the sheets in the correct sequence.

Index

Printed in the United States
71131LV00004B/175-320

9 780471 143420